Lecture Notes in Chemistry

Volume 115

The series Lecture Notes in Chemistry (LNC), reports new developments in chemistry and molecular science - quickly and informally, but with a high quality and the explicit aim to summarize and communicate current knowledge for teaching and training purposes. Books published in this series are conceived as bridging material between advanced graduate textbooks and the forefront of research. They will serve the following purposes:

- provide an accessible introduction to the field to postgraduate students and nonspecialist researchers from related areas,
- provide a source of advanced teaching material for specialized seminars, courses and schools, and
- be readily accessible in print and online.

The series covers all established fields of chemistry such as analytical chemistry, organic chemistry, inorganic chemistry, physical chemistry including electrochemistry, theoretical and computational chemistry, industrial chemistry, and catalysis. It is also a particularly suitable forum for volumes addressing the interfaces of chemistry with other disciplines, such as biology, medicine, physics, engineering, materials science including polymer and nanoscience, or earth and environmental science.

Both authored and edited volumes will be considered for publication. Edited volumes should however consist of a very limited number of contributions only. Proceedings will not be considered for LNC.

The year 2010 marked the relaunch of LNC.

Nagahiro Saito
Editor

Solution Plasma Chemistry

Springer

Editor
Nagahiro Saito
Nagoya University
Nagoya, Aichi, Japan

ISSN 0342-4901 ISSN 2192-6603 (electronic)
Lecture Notes in Chemistry
ISBN 978-981-95-0965-2 ISBN 978-981-95-0966-9 (eBook)
https://doi.org/10.1007/978-981-95-0966-9

This Springer imprint is published by the registered company Springer Nature Singapore Pte Ltd.
The registered company address is: 152 Beach Road, #21-01/04 Gateway East, Singapore 189721, Singapore

Preface

This book is the culmination of broad and extensive collaboration with many colleagues. Chapters "What Are Solution Plasmas?"–"Characteristics of Solution Plasmas" were co-authored with Dr. Maria Antoaneta Bratescu, former Specially Appointed Professor at Nagoya University. Chapters "Electrochemistry in Solution Plasma"–"Materials Synthesis Using the Solution Plasma Reaction" reflect joint research with Dr. Chayanapat Chokratanaruean, Associate Professor at Shinshu University, and Chapter "Toward Industrial Solution Plasma Process" was written in collaboration with Prof. Tatsuru Shirafuji of Osaka Metropolitan University. I extend my deepest gratitude to all co-authors for their devoted efforts, profound insights, and generous cooperation throughout this research journey.

The inception of this book traces back to the concept of "non-equilibrium plasma in liquids," which began to take shape in my mind between 2000 and 2003. During my doctoral studies, I investigated reaction mechanisms in chemistry, but it was only after moving to Nagoya University that I was able to fully immerse myself in the world of plasma science. I was always conscious of the gap that existed between chemistry and plasma research, and it was precisely this divide that stimulated my curiosity and willingness to take on new challenges.

Around that time, I encountered a paper on the Miller experiment and was deeply struck by its bold attempt to induce novel molecular transformations using non-equilibrium plasma. Although its significance for the origin of life has since been reconsidered, the sense of excitement I felt at that time has never faded. I was particularly inspired by the fact that the experiment was led by a graduate student, someone of my own generation.

These encounters gave me powerful motivation to venture into uncharted territory. I became determined to develop cold plasma operating in liquids and to create a new tool for chemical synthesis. However, generating stable plasma in liquids was by no means an easy task. In the initial experiments, arc and streamer discharges occurred frequently, each posing its own chemical challenges.

Through persistent effort and repeated trial and error, I eventually succeeded in realizing a plasma state at the solution–plasma interface that resembled glow discharge. In this state, electrons could move freely, making new reactions possible.

I discovered that by optimizing the combination of solvent and solute, a wide variety of chemical transformations could be promoted, and I named this technique "solution plasma." What began as research on water discharge and sterilization rapidly expanded to include the synthesis of nanomaterials such as graphene, as well as organic-phase reactions with ever-widening applications.

Such progress would not have been possible without the passionate students and many collaborators who joined along the way. Their enthusiasm and creativity have driven the expansion of solution plasma research into areas such as natural polymers, catalysis, and surface modification. I wish to express my heartfelt gratitude to everyone who contributed to these developments.

There were three major turning points and sources of support that paved the way for this research. First, Prof. Osamu Yanagida, Professor Emeritus of Nagoya University, provided invaluable encouragement at the early stage of the project. Second, Prof. Yasuhiro Horiike, former Professor at the University of Tokyo, supported the establishment of our research base through JST CREST funding. Third, Prof. Hisashi Yamamoto, Professor Emeritus of the University of Chicago, enabled a shift from a physics-centered to a chemistry-centered approach by including this project in the JST CREST "Molecular Technology" initiative. I am deeply grateful for the guidance and support of these mentors.

The idea for this book originated when I received a proposal from Springer to write a textbook during the course of the Molecular Technology project. Solution plasma chemistry remains a developing field with many unknowns, and I initially hesitated to take on this challenge. However, hoping that this book would serve future researchers and provide a record of what has been learned so far, I resolved to undertake the writing. Due in part to my own delays, the completion of this book took more than ten years, and I am sincerely grateful to the publisher for their understanding and support throughout the process.

Our research has demonstrated that solution plasma is fundamentally different from conventional gas-phase plasma and ordinary liquid-phase reactions. By generating plasma within liquids, it becomes possible to design entirely new reaction environments and precisely control the formation of reactive species, thereby achieving high selectivity and energy efficiency. We have made advances in a wide range of applications, including the synthesis of graphene and nitrogen-doped graphene from aromatic compounds, reduction of metal ions, and new strategies for doping photocatalytic materials. In recent years, interdisciplinary approaches have led to further progress, such as elucidating physical and chemical parameters at the plasma–solution interface, identifying novel reaction intermediates, and incorporating AI into the analysis of reaction mechanisms. The integration of diverse research fields, along with the aspiration to reconstruct basic chemistry for a sustainable and decarbonized society, continues to open up new possibilities for this science.

Even as I write these lines, the science and applications of solution plasma are evolving at a remarkable pace. I look forward with great anticipation to the achievements that will be made by the next generation of researchers and students. Above all, I hope that this book will stimulate curiosity, foster the spirit of inquiry, and serve as a guide not only for young scientists but also for children who have yet to embark on their journey in science, lighting the path for those who will shape the future.

In my Minamichita home, where tides and laughter gently rhyme

Nagoya, Japan
June 2025

Nagahiro Saito

Contents

What Are Solution Plasmas?

Nagahiro Saito and Maria Antoaneta Bratescu

Abstract Solution plasmas usually refer the plasmas in liquids. Either plasma in gas or in liquid, they are electrical discharge. Therefore, they are believed to share fundamentals, including molecular interaction (e.g., attractive and repulsive forces between molecules), breakdown, propagation, and formation of plasma layers, referring to plasma sheath. However, due to density of media, plasma in liquid has the altered characteristics, compared to that of plasma in gas. Accordingly, the clear understanding of plasmas in liquid, like solution plasma, should be explained and applied, aiming to maximize the benefit of plasma technology. In this chapter, the general explanation relating to the definitions and basic fundamental relating to SP is provided.

1 Introduction

In 1887, British physicist J.J. Thomson conducted experiments with cathode ray tubes and showed that all atoms contain tiny negatively charged subatomic particles or electrons (Fig. 1). These tiny negatively charged particles were found to co-exist with other particles, including positively charged and neutral particles. The collection of these particles was first dubbed "plasma" by Irving Langmuir in 1928. In general, plasma is known as the fourth state of matter or ionized gas.

What is ionized gas? Ice (solid) transits to water (liquid) and water vapor (gas) with increasing temperature, as shown in Fig. 2. These are called the three states of matter. As temperature continuously rises, collisions of atoms and molecules in the gas phase become more frequent and intense. At a specific temperature, at least one electron leaves an atom or molecule, transforming the atom or molecule into positively changed ions (Fig. 3). The process is called "ionization," resulting in the

N. Saito (✉)
Institute for Innovation for Future Society, Department of Chemical Systems Engineering, Graduate School of Engineering, Nagoya University, Nagoya, Japan
e-mail: saito.nagahiro.z7@f.mail.nagoya-u.ac.jp

M. A. Bratescu
Institute for Innovation for Future Society, Nagoya University, Nagoya, Japan

N. Saito (ed.), *Solution Plasma Chemistry*, Lecture Notes in Chemistry 115,
https://doi.org/10.1007/978-981-95-0966-9_1

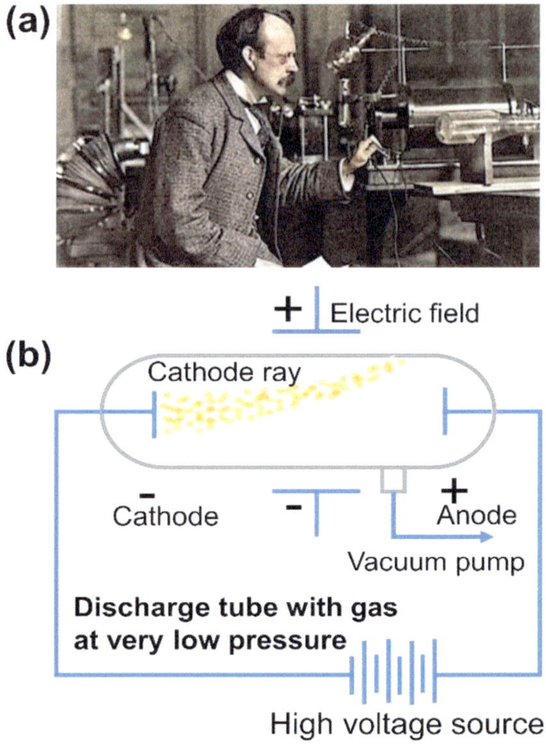

Fig. 1 **a** J. J. Thomson produced a visible beam in a cathode ray tube; **b** Deflection of the cathode ray towards the positive plate while traveling to the anode in a vacuum tube

generation of the ionized gas. However, ionized gas can be called plasma when electron density is balanced with positively charged ions. In other words, the total number of positively and negatively charged ions in plasma is generally equal, called "quasi-electrically neutral." Moreover, it should contain enough electrically charged particles, which can influence its electrical behavior.

In plasma, at least one of the electrically charged particles, e.g., electrons, moves freely through space without being restricted by other groups and repeatedly collides with surrounding particles. This phenomenon is common to the behavior of free electrons in metals. Therefore, we can say that "plasma is conductive" in the same manner that metals are conductive. Sometimes, the electrically charged particles can make plasma electrically conductive more than gold and copper. However, plasma differs from metals in one significant aspect. Not only electrons but also positively and negatively charged ions can easily move in plasma. Accordingly, plasma is electrically conductive, internally interactive, and significantly responsive to the electromagnetic field. These positively and negatively charged ions occupy spaces in plasma. In other words, the spatial distribution of charged particles tends to occur. Spatial distribution normally relates to densities, concentrations, and patterns, such as the distribution of electron density and temperature. This phenomenon is the same as the formation

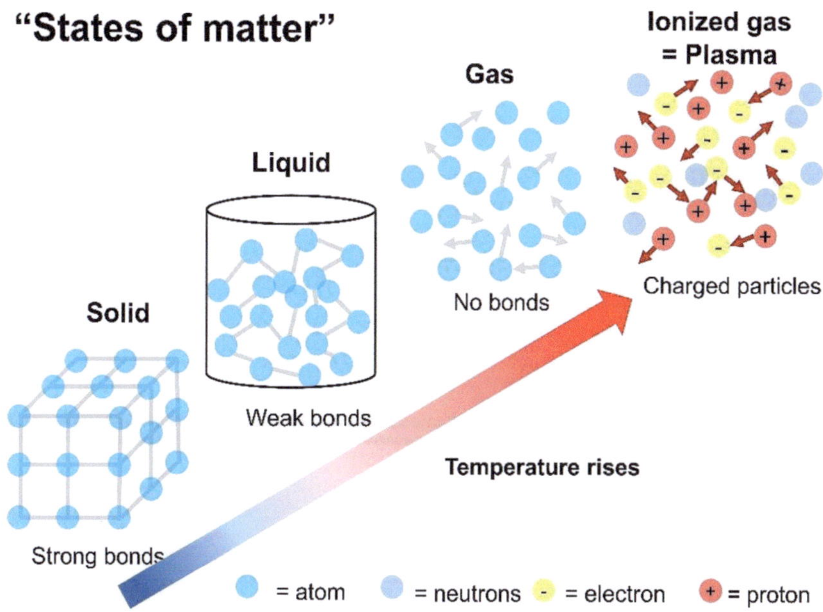

Fig. 2 States of matter diagram, including solid, liquid, gas, and plasma

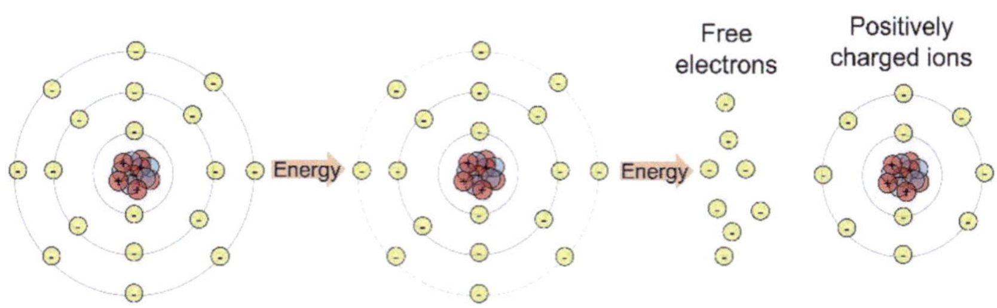

Fig. 3 Electron configuration of Argon (Ar), containing 18 electrons, 18 protons, and 22 neutrons; after adding energy, such as heat, an electron leaves an atom to form positively charged ions

of spatial distribution that occurs between two electrodes placed in an electrolyte solution, which has been studied in electrochemistry.

The positively and negatively charged ions generated in the plasma determine the electrical characteristics of the plasma, and the types of positive and negative ions greatly depend on the raw material gas species. For example, if the raw material gas is only argon, Ar^+ ions (positive ions) are generated, whereas negative ions are not. On the other hand, when oxygen is added to the raw material gas, O_2^- ions (negative ions) are also generated. The mobility of these ions differs greatly depending on the characteristics of the ion species, such as their charges and masses. In plasma, particle motion is mainly caused by interactions between charged particles, which

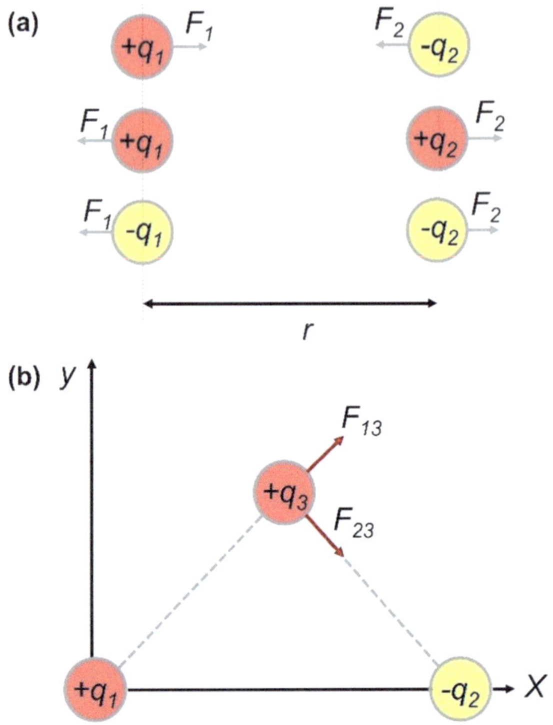

Fig. 4 Coulomb force is the electrostatic force between two point charges (q_1,q_2) is directly proportional to the product of the magnitudes of charges and inversely proportional to the square of the distance (r) between them; $|F| = K\frac{|q_1||q_2|}{r^2}$, here, K or k_e is Coulomb's constant ($k_e \approx 8.988 \times 10^9$ N·m^2·C^{-2}). The force is along the straight line joining the two charges. If the charges have the same sign, the electrostatic force between them is repulsive; if they have different signs, the force between them is attractive

are exerted by the attraction and repulsion forces, i.e., Coulomb forces, as revealed in Fig. 4. Due to these Coulomb forces, plasmas have wide wavelengths range and finite frequencies range. As a result, the spatial distribution of charged particles in the plasma varies with different source gases.

Plasma naturally exists, such as in the sun, stars, and lightning. On the other hand, plasma can also be induced by applying external electric fields using the power supply: it sometimes is called "artificial plasma". When a voltage is applied, the motions of particles between electrodes occur, resulting in the collision of particles. One particle travels through a distance in plasma to collide with another particle. The distance, on average, is called the "mean free path," which can influence the properties of the plasma. The mean free path usually depends on applied pressure. For example, a high vacuum (10^{-3}–10^{-7} mbar) can result in the mean free path ranging from centimeters to kilometers, whereas that at atmospheric pressure (10^3 mbar) is in the order of tens of nanometers (Ollegott et al. 2020). A shorter mean free path leads to a higher collision frequency. Moreover, not only the applied pressure but also the voltage is important for igniting and sustaining the plasma. The breakdown voltage required to ignite the plasma follows Paschen's law for simple planar metal electrodes in the gas phase, as shown in Fig. 5a. The breakdown voltage was found to depend on the pressure (p) and the gap distance (d) between two electrodes and differ

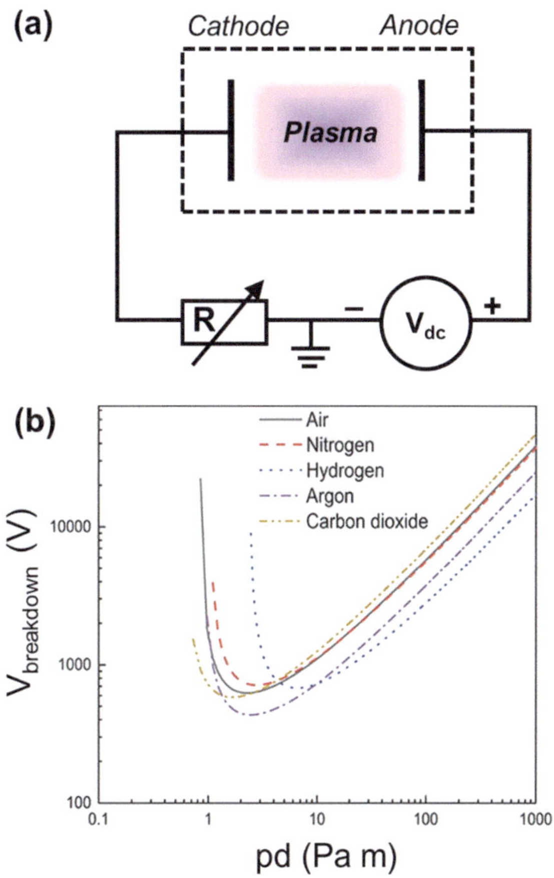

Fig. 5 **a** Schematic diagram of a DC discharge between parallel plate electrodes and **b** Paschen curves, showing the breakdown voltage as a function of pd for different gases (Reproduced from (Ollegott et al. 2020))

for different types of gases. This is because different gases have different ionization potentials and mean free paths. Paschen curves for several gases are shown in Fig. 5b.

In the electric breakdown process, primary (free) electrons are accelerated in the electric field and collide with gas molecules. Then, electrons, bounded in gas molecules, are ejected by impact ionization, leading to the formation of positively charged ions. Afterward, the ejected electrons are repeatedly accelerated by the external electric field and further collide with gas molecules. Multiple repetitions of the process continuously occur, leading to the generation of electron avalanches, as illustrated in Fig. 6. In addition, when the positively charged ions are accelerated and collide with the cathode material, the emission of secondary electrons happens and initiates additional avalanches. After the breakdown, when the constant energy is input, the propagation, i.e., plasma, rapidly expands to obtain a steady stream and occurs to form the plasma between electrodes. Meantime, the continuous emission of secondary electrons from the cathode helps sustain the plasma.

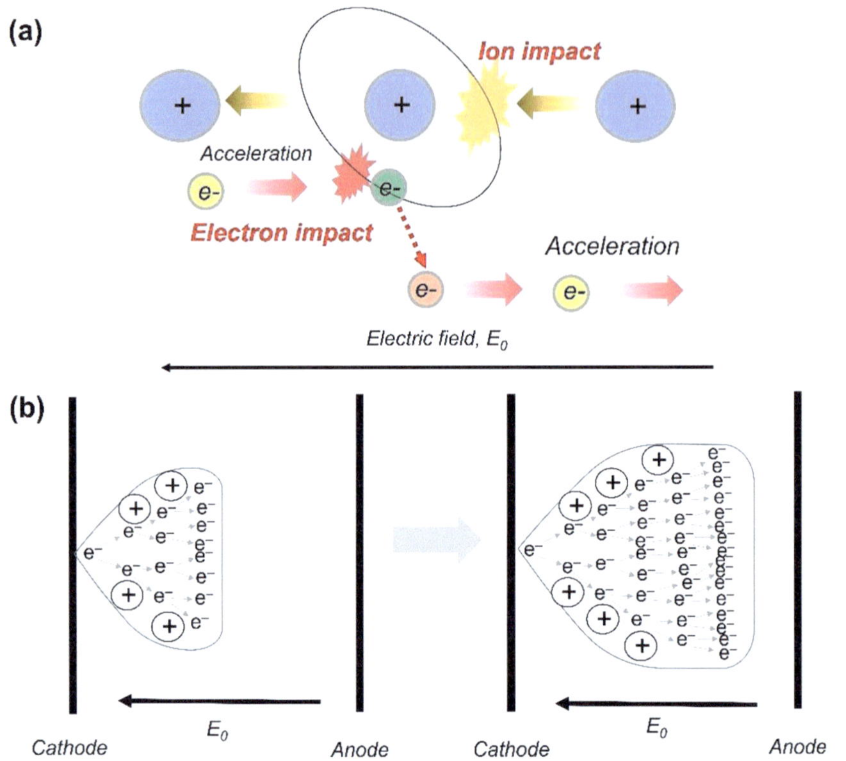

Fig. 6 **a** impact ionization after applying voltage and **b** formation of electron avalanche

When an external electric field is applied, particles in the plasma move (charge transport), forming a spatial distribution of charges. Electrons are mostly depleted at the interface between plasma and electrodes. Subsequently, a region dominated by positive ions and neutrals is formed. This phenomenon results in the formation of an electric double layer in the plasma, as shown in Fig. 7, which usually exhibits a dark region near the electrode. The electric double layer is known as the "plasma sheath" and is recognized as an important region that determines the reaction mechanism. In addition, the direction of charge transfer causes an electric current flow, which generates a magnetic field. In the plasma sheath, the potential (V) gradient is steep, which significantly affects the motion of each particle, i.e., the form of the reaction.

As mentioned above, particles in a plasma are in motion. The particle motion relates to the energy of particles and their degree of freedom (e.g., translational, rotational, and vibrational motions), which determine the temperature in plasma. Temperature is the variable of chemical reactions and strongly influences the reactions that occur in the plasma or at its boundaries. Plasma is a multi-component system consisting of electrons and negatively and positively charged particles (heavy particles); electron temperature (T_e), ion temperature (T_{ion}), and gas temperature (T_g)

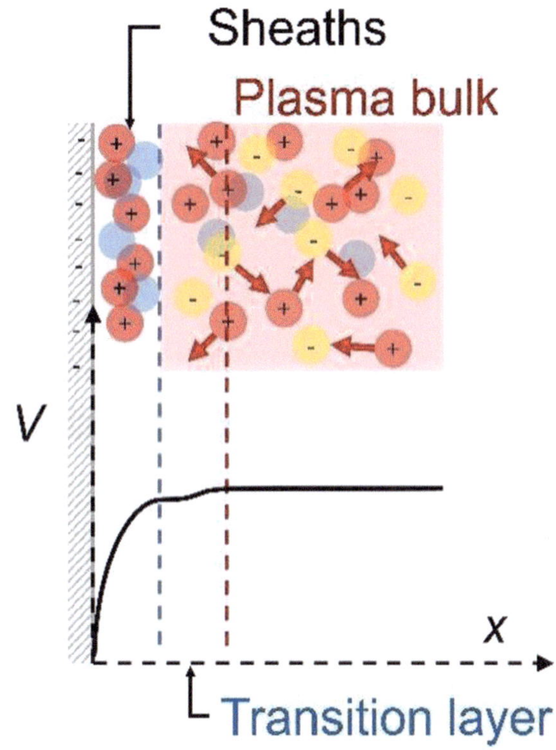

Fig. 7 Discharge behavior and formation of plasma sheath, low mass warm electrons quickly drain to the wall, forming sheaths, while ions accelerated to walls and bulk plasma is quasi-neutral

are usually mentioned for the classification of plasmas. Plasmas are classified into (i) equilibrium and (ii) nonequilibrium plasmas in terms of thermal equilibrium and are usually referred to as thermal or non-thermal plasmas. If the temperatures of each particle are all equal, the state is said to be in "thermodynamic equilibrium." Conversely, if the temperatures of the particles are not all equal, the state is said to be in "thermodynamic nonequilibrium." In the case of a typical plasma, if the pressure is high, the plasma is in equilibrium. On the other hand, at lower pressures, the plasma is nonequilibrium. To be specific (Fig. 8), equilibrium plasma exhibits a local thermodynamic equilibrium (LTE) between electrons and the other ions ($T_e \approx T_{ion} \approx T_g \approx 4 \times 10^3$ to 20×10^3 K) and can be generated at a pressure higher than atmospheric pressure ($\geq 10^5$ Pa). Contrastingly, a nonequilibrium plasma can usually be obtained at a pressure much lower than the atmospheric pressure. In the nonequilibrium plasma, the T_e is usually at 10^4 K, whereas the temperature of other ions is close to the ambient temperature ($T_e >> T_{ion} \approx T_g \approx 300$–$400$ K).

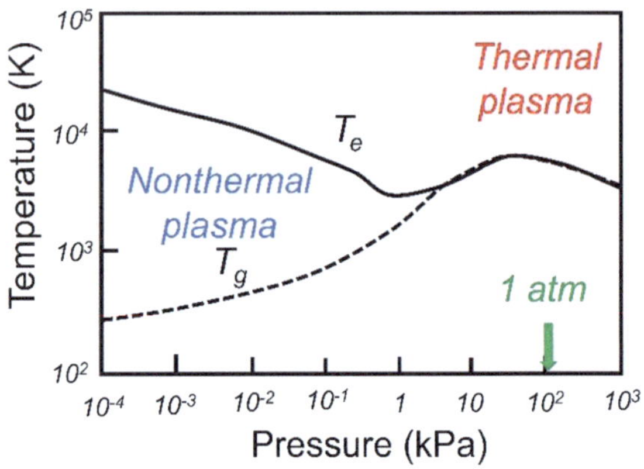

Fig. 8 Relationship of electron temperature (T_e) and gas temperature (T_g), referring to the temperature of ions and pressure. (Boulos et al. 1994), with permission from Plenum Press; Book: Thermal plasmas: fundamentals and applications)

2 Non-Equilibrium Plasma in Liquid: Solution Plasma

Here, Fig. 9 shows the transition from thermal non-equilibrium to thermal equilibrium, which depended on the temperature of electrons and ions at different pressures and temperatures of the discharge environment. For the past 40 years, nonequilibrium plasma under atmospheric pressure has been studied. Nonequilibrium plasmas in liquids are an extension of this research. For some nonequilibrium plasmas in liquids, the term "Solution Plasma" was proposed by Takai in 2008. Since solution environments can contain a wide variety of solutes and solvents, plasmas occurring in such environments are collectively referred to as solution plasmas.

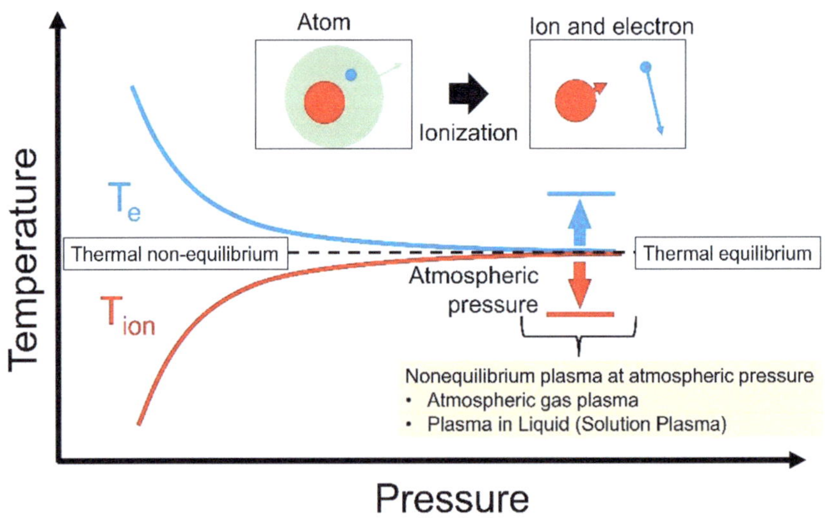

Fig. 9 Electron and ion temperatures in nonequilibrium plasmas under different pressures

2.1 *Solution Plasma Formation and Features*

The formation of plasmas in liquids usually includes several physical features, such as breakdown and propagation, which are different from plasmas in gases. The reason is that the density of the liquid surrounding the plasma is extremely high compared to that of the gas. Therefore, a system with a large chemical reaction rate can be obtained compared to ordinary nonequilibrium atmospheric pressure plasma generated in gas. On the other hand, the high density of liquid hinders the formation of plasmas unless an electric field with several orders of magnitude higher than that of ordinary nonequilibrium atmospheric pressure plasma is provided. Moreover, because of the pressure and density of a liquid, a transition to a thermal equilibrium state, e.g., arc plasma, can occur and lead to extremely high temperatures (4×10^3 to 20×10^3 K), as revealed in Fig. 10. Recently, it has been possible to form nonequilibrium plasmas in liquids by using bipolar square waves. The solution plasma with the typical pin-to-pin electrode and bipolar pulsed power supply, as shown in Fig. 11, can induce nonequilibrium plasmas, e.g., glow discharge. In general, electrons in plasmas can be instantaneously accelerated by applying electrical energy, while the acceleration of ions is much slower due to their masses. When the voltage is continuously applied, the energy is accumulated, resulting in the intense acceleration of the ions; thus, the plasma reaches a thermal equilibrium state. Meanwhile, using a bipolar pulsed power supply in the solution plasma can help to suppress the acceleration of ions and prevent temperature rises.

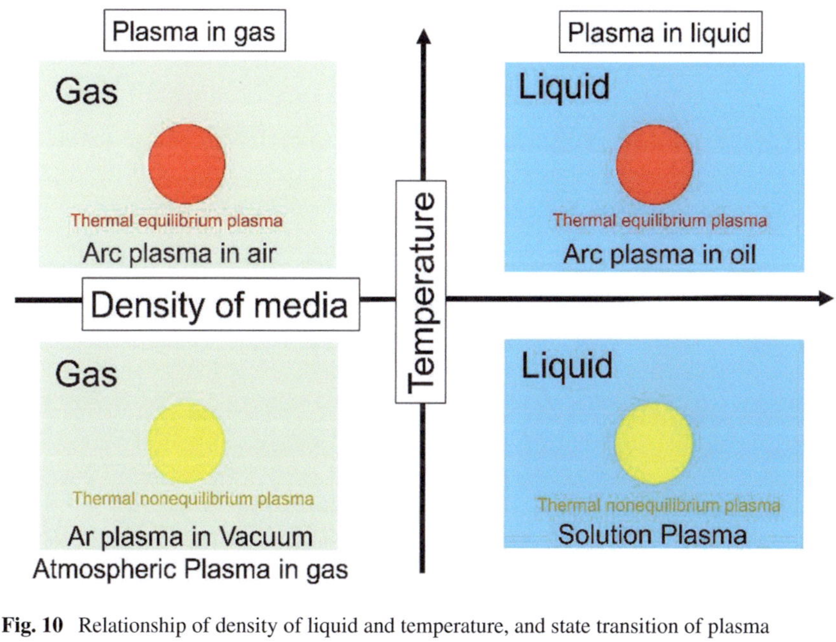

Fig. 10 Relationship of density of liquid and temperature, and state transition of plasma

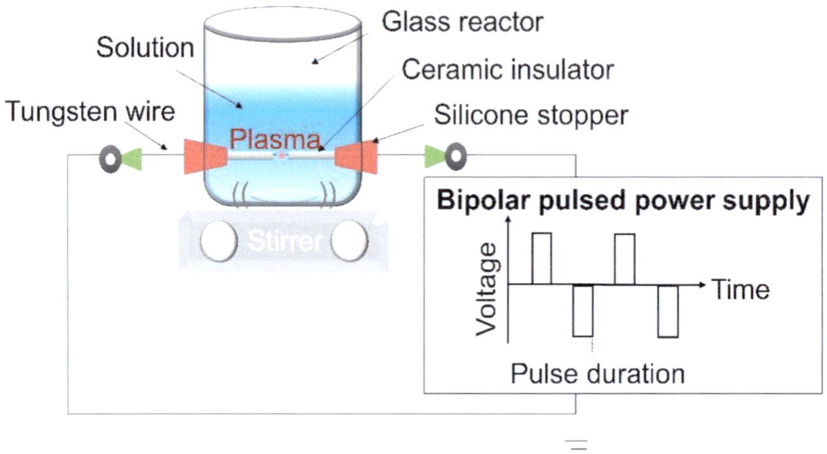

Fig. 11 Typical solution plasma setup

Figure 12 Summarizes the comparison between the arc and the glow plasmas. For arc plasma, a chemical reaction process is generally an energy exchange through collisions. As a result, the reaction has a high speed and high mass transfer. However, there is also a high degree of electrode destruction by forming a hot spot on the electrode surface. In contrast, the glow plasma can induce a chemical reaction process via electron exchange through collisions and the formation of secondary electrons. The reaction process can be kinetically controlled (Chae et al. 2019). As a result, the reaction can reduce electrode damage and increase the selectivity of the reaction

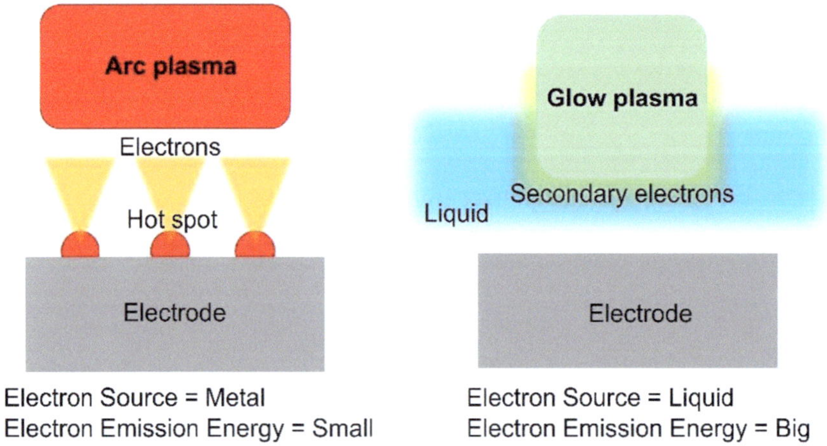

Fig. 12 Comparison between the arc and the glow plasmas in liquid

2.2 Solution Plasma Reaction Field and Chemistry

The components of solution plasma consist of plasma located at the center and surrounded by a gas phase (bubbles). The outer layer of the gas phase is surrounded by a liquid phase. Therefore, there are two interfaces, plasma/gas and gas/liquid (Fig. 13a). The previous studies made the spectroscopic measurements of reactive species in the solution plasma, as shown in Fig. 13b. As a result, plasma chemical reaction fields were revealed to be (i) at the center of plasma and (ii) at the gas/liquid interface. The reaction at the center of plasma mainly occurs through collisional processes, e.g., dissociation, excitation, and ionization.

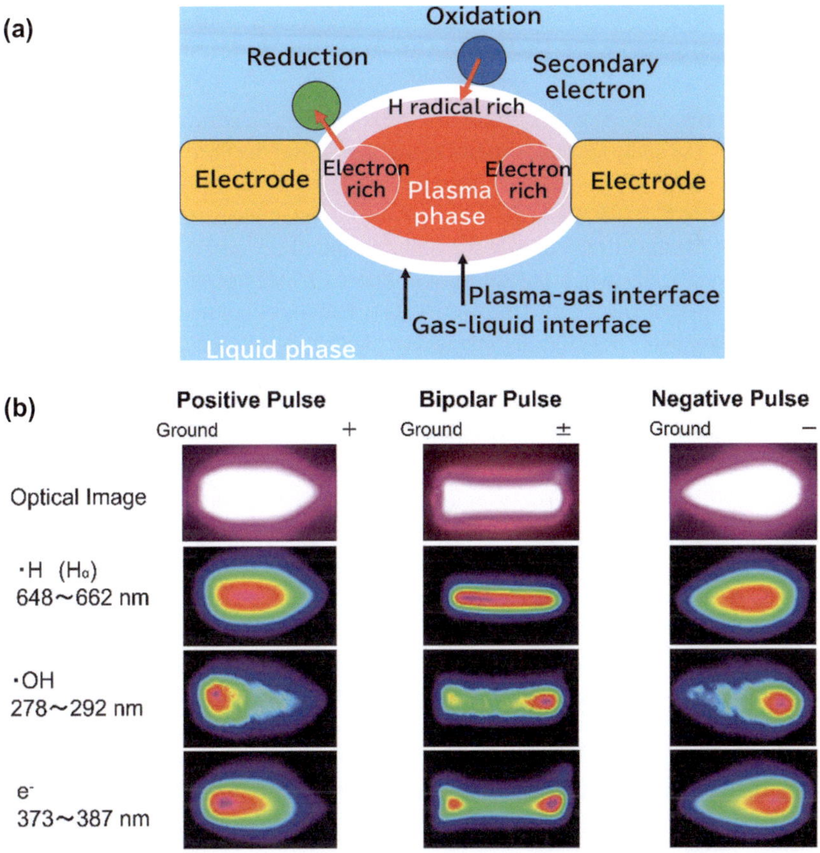

Fig. 13 **a** schematic diagram of plasma between pin-to-pin electrodes under liquid phase and **b** chemical contrast images of active substances in aqueous discharge plasma: reference microscopic images; hydrogen radical (bandpass wavelength 648–662 nm); hydroxyl radical (bandpass wavelength 278–292 nm); electrons (bandpass wavelength 373–387 nm) (Reproduced from (Yui and Banno 2017))

Meanwhile, the reaction at the interfaces is dominated by electron accepting and donating, i.e., oxidation and reduction, referring to electrochemistry. Furthermore, a unique feature of liquids is that a mixture of at least two or more solutions can be used instead of a single liquid. Here, various solutions from different chemical disciplines can be used, such as solution chemistry. In other words, a wide variety of reaction systems can be provided. Because it is a plasma in solution under atmospheric pressure, it is extremely compatible with conventional chemical engineering plant technology. For this reason, unlike plasma in the semiconductor field or plasma in the surface treatment field, it can be called chemical engineering plasma, which is suitable for handling by chemists.

2.3 Foundations for Building the Solution Plasma Field

Plasma chemistry in solution encompasses three major scientific disciplines. The first is plasma physics, the second is solution chemistry, and the third is electrochemistry. Plasma physics allows us to understand the solution plasma itself, even if it is plasma in solution, because it does not deviate from the general concept of plasma. In solution plasma, the chemical reactions at the solution interface are induced by the plasma change dramatically when the solution is altered by adjusting the combination of solute and solvent. This is the most important feature of solution plasma. The third is electrochemistry. In solution plasma, the plasma in solution and the solution can be regarded as two electrochemical electrodes, respectively. The potential difference between them affects oxidation–reduction reactions, and there are several reactions that are similar to electrochemistry. For example, platinum chloride and gold chloride are reduced to the metal platinum and gold by the plasma in solution, while copper sulfate solutions are reduced only under certain solution conditions. This is consistent with the standard electrode potential of the metal. However, solution plasma in water produces hydrogen peroxide but not ozone, even when sufficient voltage, such as 1000 V, is applied to the electrodes. Thus, it differs from electrochemistry in many ways. At first glance, they do not match standard electrode potentials. However, when it is reconsidered as electrochemistry with two poles, solution and plasma, it agrees with the explanation from the standard electrode potentials. This potential effect contributes significantly to the reaction selectivity of the solution plasma. Chemical engineering is also an important area of engineering research with the industrialization of solution plasmas in mind. This is a relevant area because of the medium and morphology properties of plasmas in solution, which are not present in gas-phase plasma processes.

This book will follow plasma physics, solution chemistry, electrochemistry, plasma applications, and industrialization.

References

Boulos MI, Fauchais P, Pfender E (1994) Thermal plasmas: fundamentals and applications, vol 1. Springer Science & Business Media, New York

Chae S, Panomsuwan G, Bratescu MA, Teshima K, Saito N (2019) P-Type doping of graphene with cationic nitrogen. ACS Appl Nano Mater 2(3):1350–1355

Ollegott K, Wirth P, Oberste-Beulmann C, Awakowicz P, Muhler M (2020) Fundamental properties and applications of dielectric barrier discharges in plasma-catalytic processes at atmospheric pressure. Chem Ing Tec 92(10):1542–1558

Saito N, Bratescu MA, Hashimi K (2017) Solution plasma: a new reaction field for nanomaterials synthesis. Japanese J Appl Phys **57**(1):0102A4

Yui H, Banno M (2017) Microspectroscopic imaging of solution plasma: how do its physical properties and chemical species evolve in atmospheric-pressure water vapor bubbles? Japanese J Appl Phys 57(1):0102A1

Basics of Plasma Physics

Nagahiro Saito and Maria Antoaneta Bratescu

Abstract Solution Plasma (SP) is an electrical discharge at atmospheric pressure in the gas formed from a surrounding liquid. Water, aqueous solution, organic compound, organic solution, suspension, liquid nitrogen, supercritical fluids, or ionic liquids can be the solution in the SP. The SP is a cold and non-local Thermodynamic Equilibrium plasma where the electron temperature is much greater than the gas temperature. The SP operates in a glow discharge regime. The processes inside SP are named Solution Plasma Processes (SPP). SPP combines the physics of atmospheric pressure discharges, the thermodynamic properties of liquids, and the physical and chemical reactions at the interfaces: plasma gas-phase and solution, plasma gas-phase and electrodes surface, and solution-electrodes interface. Four species of particles exist in a plasma gas phase: electrons, ions, neutrals, and radicals. The species densities and temperatures characterize the plasma. This chapter is focused on the formation of the SP and the diagnostics methods using electrical and optical procedures. The electrical characteristics of the SP mean the measurement of the current, voltage, and electron temperature. The optical diagnostics methods like emission and absorption spectroscopy measure the electron and particle densities in the SP gas and liquid phases.

1 Introduction

Plasma physics is a chapter of physics that experimentally and theoretically studies the plasma, a state of matter distinct from solids, liquids, and gases. Plasma is considered the fourth state of matter and is characterized by its unique properties (Fig. 2 in Chap. "What are Solution Plasmas?"). Plasma is an electrically conducting medium in which roughly equal numbers of positively and negatively charged particles are

N. Saito (✉)
Department of Chemical Systems Engineering, Graduate School of Engineering, Institute for Innovation for Future Society, Nagoya University, Furocho, Chikusa, Nagoya, Japan
e-mail: saito.nagahiro.z7@f.mail.nagoya-u.ac.jp

M. A. Bratescu
Institute for Innovation for Future Society, Nagoya University, Furocho, Chikusa, Nagoya, Japan

N. Saito (ed.), *Solution Plasma Chemistry*, Lecture Notes in Chemistry 115,
https://doi.org/10.1007/978-981-95-0966-9_2

produced when the atoms or molecules in a gas become ionized. In a plasma, some or all of the electrons are stripped from their parent atoms and molecules, resulting in a mixture of positively charged ions, free electrons, and neutrals, including radicals and excited states of atoms and molecusles. In some plasmas, electrons can attach to an atom or molecule, creating negative ions. Some key features of plasmas and the main areas of study are ionization, charged particles, and collective behavior. A plasma is formed when energy is injected into a gas, causing the atoms or molecules to lose electrons. This process is called ionization. Plasma consists of charged particles, positive and negative ions, and free electrons that interact with each other through electromagnetic forces. Plasma physics often works with the collective behavior of charged particles, where the motion of one particle can influence the motion of another particle due to the long-range electromagnetic forces.

2 Kinetics of Moving Neutrals and Charges

2.1 Neutral Particles

To understand plasma parameters, the knowledge of the simple kinetics of moving neutral and charged particles, including their collisions, is necessary. The kinetic description of a stationary gas deals with many neutral particles (atoms or molecules) with a wide velocity distribution moving in all directions. The velocity distribution is given by an average energy $\langle \varepsilon \rangle$ connected to the kinetic temperature or thermal energy density, $k_B\ T$, where k_B is the Boltzmann constant ($k_B = 1.38 \times 10^{-23} \mathrm{JK}^{-1}$) and T is temperature. Assuming thermal equilibrium, the gas pressure, p, represents the thermal energy density multiplied by the gas density, n_g:

$$p = n_g k_B T \tag{1}$$

The most probable velocity of a moving neutral particle (v) with an atomic mass m is $v = \sqrt{\frac{2k_B T}{m}}$. The average velocity is $v = \sqrt{\frac{8k_B T}{\pi m}}$. The flux of particles is $\Gamma = n_g \sqrt{\frac{k_B T}{2\pi m}}$. The mean free path, which is the distance that a neutral particle travels between two consecutive collisions, λ, is

$$\lambda = \frac{1}{4\pi r_p^2 n_g} \tag{2}$$

where r_p is the particle radius. The frequency of the collisions between the gas particles v is $\nu = \frac{\langle v \rangle}{\lambda} \lim_{x \to \infty}$.

2.2 Charged Particles

The introduction of charged particles results in a few changes. First, the charge produces an electric field through which it exerts forces on any other charges. The force is in inverse square proportion to the separation distance (r^{-2}). Second, an applied external electric field exerts a force on the charged particle. Third, the charged particles also move in a magnetic field. The Lorentz force combines the electric and magnetic effects for a charged particle $\vec{F} = q\left(\vec{E} + v \times \vec{B}\right)$, where q is the particle charge, $\vec{F}$ is the force on the charged particle due to the electric field $\vec{E}$ and the magnetic field $\vec{B}$.

An electron or a charged particle gains kinetic energy in the electric field produced by a voltage V. The kinetic energy is $\frac{1}{2}mv_d^2 = eV$, where v_d is the drift velocity, and e is the elementary charge. The charged particle moves a distance x_1 along the electric field E, gaining the energy: $Ex_1 = \int_0^{x_1} eEdx$. The elastic collisions randomize the motion of the particles, including charged particles, resulting in the average velocity,$\langle v \rangle$.

2.3 Collisions and Cross-Section

In gases with neutral and charged particles, the most frequent collision (dominant interaction) is between pairs of particles (binary collisions). When particles collide, the momentum and the energy must be conserved. There are three clear types of collisions.

(1) The elastic collision occurs when the total momentum is distributed between particles and the total kinetic energy is conserved, e.g.

$$e_{fast} + A_{slow} \rightarrow e_{less\ fast} + A_{less\ slow}$$

where A represents the particle.

(2) Inelastic collisions also occur when the momentum is redistributed between particles. However, during the inelastic collision, a fraction of the initial kinetic energy is transferred into internal energy in one or more particles, i.e., the particle A is excited (A*) or ionized (A^+).

$$e_{fast} + A \rightarrow e_{slower} + A^* \text{ or } e_{fast} + A \rightarrow e_{slower} + A^+ + e_{slow}$$

(3) Superelastic collisions take place when the total kinetic energy increases after collision. The momentum is conserved, but the internal energy of one of the particles entering the collision is transferred into kinetic energy.

$$A^*_{slow} + B_{slow} \rightarrow A_{faster} + B_{faster}$$

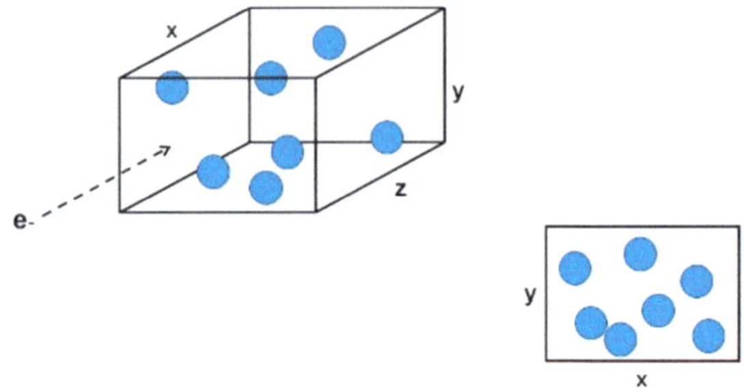

Fig. 1 The cross-section of an electron colliding with atoms

Let us consider an electron moving through many stationary atoms (Fig. 1). We consider only elastic collisions and assume that the target atoms are hard spheres while the electron is treated as a point mass. The effects arising from the charge are not included. The question is how far the electron can be expected to penetrate before a collision.

The number of target atoms in a cube xyz is $n_g xyz$. Each atom presents across-section, $\sigma = \pi r_p^2$ obscuring the electron path. Looking at the surface xy the total area blocked by atoms will be $n_g xyz\sigma$. When the cube extends as far as one mean free path (λ), virtually the entire surface xy will be obscured, so the electron is very likely to have a collision. Consequently:

$$\left(n_g xy\lambda\right)\sigma = xy \text{ and } \lambda = \frac{1}{n_g\sigma}. \tag{3}$$

The frequency of collisions will be $\upsilon = vn_g\sigma$.

3 Plasma Parameters

In plasma physics, various parameters are used to characterize the properties of a plasma. Basic plasma parameters are related to the kinetics of neutral and charged particles moving and the interactions among particles. The parameters provide information about the temperature, density, neutrality, Debye length, collision frequency, mean free paths, particle composition in plasma, external electric field, and magnetic field strength.

3.1 *Plasma Temperature*

The plasma temperature measures the average kinetic energy of the particles in the plasma. T_e, T_i, and T_n are the electron, ion, and neutral particle temperatures. The unit is Kelvin (K) or electron volts (eV) (1 K = 8.6173 × 10^{-5} eV). Based on temperature, there are two types of plasma: thermal and cold. The thermal plasmas have high temperatures of the neutral particles and ions, typically above several thousand Kelvin. Examples include plasmas in stars (10^5 to 10^6 K), fusion experiments (10^6–10^8 K), and industrial applications like plasma cutting (10^3–10^4 K). Cold or non-thermal plasmas have neutral particles and ion temperatures much lower than thermal plasmas. The temperature range of neutral particles and ions for cold plasmas would typically be approximately 300–800 K. Cold plasmas are used in technologies such as plasma TVs, fluorescent lights, and specific medical applications.

The temperatures determine the degree of ionization, the electrical conductivity, the plasma pressure, frequency, oscillations, and the thermodynamic equilibrium. Higher temperatures increase the kinetic energy of particles, making it easier for electrons to escape from atoms and molecules and create ions. As temperature rises, plasma can transit from partially to fully ionized states, changing the degree of ionization. The electrical conductivity increases with temperature because more electrons are freed from atoms and molecules, which provides more charged particles to carry the current. The plasma pressure is proportional to the temperature and density (Eq. 1), affecting plasma expansion, stability, and confinement, particularly in magnetic fields.

From the equilibrium temperature point of view, plasmas can be in thermodynamic equilibrium where the electron, ion, and neutral temperatures are approximately equal. Plasma in thermodynamic equilibrium is thermal plasma and occurs in stars and specific laboratory plasma devices such as fusion and cutting plasma. In the non-local thermodynamic equilibrium (non-LTE) plasmas, the temperatures of electrons, ions, and neutrals are unequal. Sometimes, the thermal equilibrium can be locally realized even in such plasmas. Non-LTE plasma includes the low-pressure glow, radiofrequency (RF) plasma, corona, barrier, and hollow cathode discharges at various pressures. These are commonly used in applications like plasma for medicine, pollution control, and material processing.

In the non-LTE plasma, the electron temperature is much higher than the gas temperature. The gas means atoms, molecules, their excited corresponding states, radicals, and ions. T_e can reach 20,000 K in non-LTE plasmas while T_g is slightly above room temperature. Since collisions cause a continual exchange of energy between electrons and heavy particles with a tendency to equilibrate temperatures, it is more challenging to maintain non-LTE conditions at elevated pressure with high collision rates and short mean free paths. For steady-state plasmas, the deviation from LTE can be expressed by the following formula derived from an energy balance:

$$\frac{T_e - T_g}{T_e} = \frac{m_g}{4m_e}\left(\frac{\lambda_e eE}{\frac{3}{2}k_B T_e}\right)^2 \tag{4}$$

where m_g is heavy particle mass, the notations m_e, λ_e, and e are electron mass, the electron mean free path, and elementary charge, respectively. E is the electric field. The term $\lambda_e eE$ is the amount of directed energy an electron gains along one mean free path in the direction of the electric field E. The term $\frac{3}{2}k_B T_e$ means the average thermal energy. Equation 4. indicates that large mean free paths (low pressure or density), high electric fields, and low electron energies favor deviation from the LTE conditions (Becker et al. 2005).

3.2 Plasma Density and Neutrality

The plasma density (n) is the number of charged particles (ions and electrons) per unit volume, typically expressed in particles per cubic meter (m^{-3}) or particles per cubic centimeter (cm^{-3}). Based on density, there are high-density and low-density plasmas. High-density plasmas include the dense plasmas in the stellar cores or specific experiments on inertial confinement fusion. The density can vary from 10^{31} to 10^{32} m^{-3} in the stellar cores or 10^{19}–10^{21} m^{-3} in confined fusion. Low-density plasmas include the plasma display panel (10^{15}–10^{16} m^{-3}), plasma used in medicine (10^{15}–10^{17} m^{-3}), Earth's ionosphere (10^{4}–10^{12} m^{-3}), and some laboratory plasmas used in materials processing (10^{16}–10^{18} m^{-3}).

The plasma density influences other plasma parameters. The Debye length, λ_D, which determines the range of electrostatic shielding, decreases as density increases. Higher densities mean more substantial collective effects and more effective charge screening. The collisions in the plasma depend on density. Higher density leads to more frequent collisions among the particles, which affects energy dissipation, transport properties, and the thermalization of the plasma. Collisions play a significant role in balancing LTE and non-LTE behaviors.

Plasma density also affects collective behavior. Denser plasmas exhibit stronger collective behavior due to increased particle interactions. Collective behavior leads to phenomena like wave propagation and plasma instabilities, which affect plasma uniformity. High density increases the rate of radiative processes such as bremsstrahlung, recombination, and light emission.

Charge neutrality is a necessary condition for plasma formation. It means that the number of positive ions (charged atoms or molecules that have lost an electron(s)) equals the number of negative electrons. The charge neutrality is

$$n_e = \sum_i Z_i n_i \tag{5}$$

where n_e is the electron density, n_i is the ion density of various ion species, and Z_i is the charge state (ionization level) of the ith ion species.

3.3 Debye Length

The Debye length, λ_D, is a fundamental parameter in plasma physics. It characterizes the spatial scale over which mobile charged particles can screen out the electric field within the plasma. The Debye length has the formula:

$$\lambda_D = \left(\frac{\varepsilon_0 k_B T_e}{e^2 n_e}\right)^{1/2} = 743\left(\frac{T_e}{\text{eV}}\right)^{1/2}\left(\frac{n_e}{\text{cm}^{-3}}\right)^{-1/2}\text{cm} \tag{6}$$

where ε_0 is the permittivity of free space (approximately 8.854×10^{-12} F/m), k_B is the Boltzmann constant, T_e is the electron temperature in K, n_e is the electron density (number of electrons per unit volume in m^{-3}), and e is the elementary charge ($1.602176634 \times 10^{-19}$ C). We can use T_e in eV and n_e in cm^{-3} in a practical expression, as given by the third term of Eq. 6.

The Debye length represents the distance at which the electric potential created by a single charge is significantly shielded or "screened" by the presence of opposite charges in the plasma. Beyond this distance, the net electric field created by the charge is effectively canceled out by the surrounding charged particles (Fig. 2).

Table 1 summarizes the values of electron temperature and density and the Debye length in various plasmas.

In practical terms, the Debye length is crucial in understanding and modeling phenomena such as the behavior of charged particles in plasmas, the shielding of electric fields, and determining plasma sheath thickness, which is helpful in various plasma applications (Fig. 7 in Chap. "What are Solution Plasmas?"). The Debye length is inversely proportional to the square root of the plasma electron density, indicating that a higher plasma density leads to a shorter Debye length.

An ideal plasma has many particles per Debye sphere, i.e.

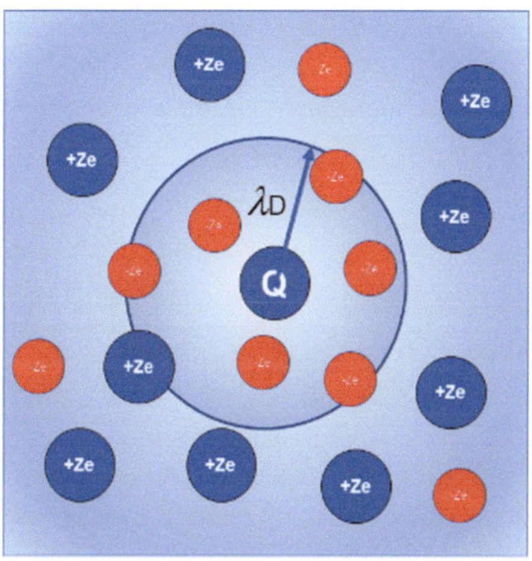

Fig. 2 The Debye length λ_D is the distance at which the surrounding ions shield the charge. The Debye sphere has a radius λ_D

Table 1 The values of electron temperature, density, and the Debye length in various plasmas

Plasma	T_e in eV	n_e in m^{-3}	λ_D in m
Laboratory plasma, low-temperature	1–10	10^{15}–10^{18}	10^{-5}–10^{-6}
Laboratory plasma, fusion plasma	10^3–10^4	10^{19}–10^{21}	10^{-7}–10^{-8}
Space plasma as the solar wind	10–100	10^6–10^8	10^2–10^3
Interstellar space	1	10^4–10^6	10^3–10^4
Stellar core	1000	10^{31}–10^{32}	10^{-14}
Supernova	10^3–10^4	10^7–10^{11}	10^{-2}–1

$$N_D \equiv n_e \frac{4\pi}{3} \lambda_D^3 \gg 1 \tag{7}$$

which is a prerequisite for collective behavior. An alternative way of expressing this condition is via the so-called plasma parameter,

$$g \equiv \frac{1}{n_e \lambda_D^3} \tag{8}$$

which is essentially the reciprocal of N_D. The assumption that g << 1 implies the dominance of collective effects over collisions between particles.

The Coulomb parameter Λ is a dimensionless parameter that accounts for the effects of long-range Coulomb interactions in plasma. The expression is:

$$\Lambda = 4\pi n \lambda_D^3 \tag{9}$$

3.4 *Collision Frequency and Mean Free Path*

Collision frequency (ν) is the frequency at which particles in the plasma collide with each other.

$$\nu = \frac{4\pi n e^4 \ln \Lambda}{m^2 v^3} \tag{10}$$

where $\langle v \rangle$ is the particle velocity, and lnΛ is the logarithm of the Coulomb parameter.

The mean free path of a moving particle in plasma is the average distance between two successive collisions. The general formula of the mean free path is $= \frac{1}{n\sigma}$, where n is the density of target particles and σ the cross-section of the collision. For Coulomb collisions (long-range interactions in plasma), the cross-section depends on the logarithm of the Coulomb parameter, which accounts for the collective effect of many weak collisions. The mean free path for Coulomb collisions λ_C becomes:

$$\lambda_C = \frac{v_r^2}{4\pi e^4 n \ln \Lambda} \tag{11}$$

where v_r is the relative velocity of the colliding particles. The physical meaning is that higher densities reduce the mean free path because there are more particles to collide. Higher temperatures increase particle velocities, influencing the mean free path through both, v_r and the logarithm of the Coulomb parameter. The types of particles and their charges affect the collision cross-section and, thus, the mean free path.

In low-temperature plasma, collision between neutral and charged particles dominates, and the mean free path ranges from 10^{-6} to 10^{-2} m. In the fusion plasma, the mean free path is 1–10^6 m due to charge-charge coulomb interactions.

3.5 Plasma Frequency

Plasma frequency (ω_p) is the frequency at which charged particles collectively oscillate.

$$\omega_p = \left(\frac{ne^2}{\varepsilon_0 m}\right)^{1/2} \text{ in rad/s.} \tag{12}$$

where n is the plasma density, e is the elementary charge, ε_0 is the vacuum permittivity, and m is the mass of the charged particle.

The plasma parameters are essential for describing the physical state of a plasma and understanding its behavior under different conditions.

3.6 Plasma Composition

The generation of electrons, ions, excited atoms, and molecules in plasma through collisions is significant for understanding plasma composition.

The species from plasma are produced through reactions involving electrons, ions, and neutrals in the plasma gas phase (Tables 2 and 3). The reactions on a surface exposed to plasma and the generation of other species from the electrode surface are shown in Table 4. The tables contain the reaction, the process, the experimental confirmation, and whether the reaction can occur in the solution plasma (SP) (Braithwaite 2000).

The main reactions of electrons in the plasma gas phase, which also occur in SP, are ionization, excitation, de-excitation, and dissociation. The most significant reactions involving ions and neutrals are ionization, excitation, Penning ionization, and dissociation. In some plasmas in molecular gases, oligomerization can occur depending on the gas precursor.

Table 2 Reactions of electrons in the plasma gas phase

Reaction	Process	Experimental confirmation	Relevant in SP
$e^- + A \rightarrow e^- + A$	Elastic collision	Thermal electrons	Yes
$e^- + A \rightarrow e^- + e^- + A^+$	Ionization	Breakdown	Yes
$e^- + A \rightarrow e^- + A^*$	Excitation		Yes
$e^- + A^* \rightarrow e^- + A + h\nu$	De-excitation	Light emission	Yes
$e^- + A^* \rightarrow e^- + e^- + A$	Two-step ionization	Efficacity of ionization	Yes
$e^- + AB \rightarrow A + B + e^-$	Dissociation Fragmentation	Light emission	Yes
$e^- + AB \rightarrow A^+ + 2e^- + B$	Dissociation and ionization		Might be
$e^- + AB \rightarrow A^- + B$	Dissociation and attachment		Might be
$e^- + A^+ + B \rightarrow A + B$	Recombination	Plasma afterglow (decay)	Might be

Table 3 Reactions in the plasma gas phase with ions and neutrals

Reaction	Process	Experimental evidence	Relevant in SP
$A^+ + B \rightarrow A + B^+$	Charge exchange	Spectra of ions	No
$A^+ + B \rightarrow A^+ + B$	Elastic collision	Spectra of ions	No
$A^+ + B \rightarrow A + B^+ + e^-$	Ionization	Ionization efficiency	Yes
$A^+ + B \rightarrow A + B^* + e^-$	Excitation	Ionization efficiency	Yes
$A + B^* \rightarrow A^+ + B + e^-$	Penning ionization	Ionization efficiency	Yes
$A^+ + BC \rightarrow A^+ + B + C$	Dissociation	Residual gas analysis	Yes
$A^+ + B + e^- \rightarrow A + B$	Recombination	Plasma decay (afterglow)	Might be
$A^\pm + B \rightarrow AB^\pm$	Oligomerization	Mass spectra	Yes
$A + B \rightarrow AB$	Oligomerization	Residual gas analysis	Yes

Table 4 Reactions on the surface

Reaction	Process	Experimental evidence	Relevant in SP
$AB + C_{solid} \rightarrow A + BC_{vapor}$	Etching	Material erosion	Might be
$AB \rightarrow A + B_{solid}$	Deposition	Thin film deposition	Might be
$A^+ + C_{metal\ electrode} + e^- \rightarrow A + C$	Sputtering	Electrode erosion	Yes
$e^- + A^+ \rightarrow A$	Recombination	Loss process	Yes
$A^* \rightarrow A$	De-excitation	Light emitted	Yes
$A^* \rightarrow A + e^-$ (from surface)	Secondary emission	Auger electrons	No
A^+(fast) $\rightarrow A + e^-$ (from surface)	Secondary emission	Auger electrons	No

In the below tables, the notation is the following: A and B are the neutral atoms, A^+ (B^+) is the positive ion, A^* (B^*) is the excited atom, AB is the molecule, e^- is the electron, C is the atom from solid material, and $h\nu$ is the photon.

The cross-section, the mean free path, and the energy of the involved particles characterize the above reactions. The cross-section of the collisions characterizes the reactions because it measures the reaction probability and efficiency. The cross-section also depends on the energy of the species involved in the reaction.

4 Plasma Generation

Plasmas can be generated through various methods, typically involving an energy input to ionize the gas and create a state where electrons are separated from atoms or molecules (Bogaerts et al. 2002). Temperature, electric field, and photons are the essential energy sources that can generate plasma.

Thermal ionization is achieved by applying heat to a gas that can raise its temperature, causing the atoms to gain enough energy to ionize and form a plasma. This method is common in stars and high-temperature plasmas. Magnetic fields are used to confine and heat a low-pressure gas, leading to ionization and plasma formation. This approach is commonly used in controlled nuclear fusion experiments, such as those conducted in tokamaks and stellarators.

The process of plasma generation by an electric field is named breakdown. Plasmas, like cold and non-LTE, obtained in the laboratory, such as electrical discharges, RF (inductive or capacity coupling) plasma, and microwave plasma, are generated by electrical breakdown. Photoionization and laser-induced plasma are generated by transferring the photon energy to a material (gas or solid) and producing ionization and plasma.

4.1 Breakdown and Townsend's Coefficients

Breakdown occurs in cold and non-LTE plasma, as does SP. As mentioned, the movement of charged particles in plasma consists of both random thermal activity and drift caused by an electric field (E). Let us suppose a gas between two metal plates–electrodes. Before the breakdown, there are few electrons in the gas, which interact a little with the gas. An existing electron is accelerated and gains energy by applying a voltage between the two electrode plates. After one mean free path of ionization collision (λ_i) with a neutral particle, the electron produces an ion and another electron. The newly created and primary electrons are accelerated in the electric field, generating new ion–electron pairs. Therefore, the number of ions and electrons exponentially grows (multiplies), generating plasma.

The first Townsend's coefficient (α) represents the number of ionization collisions an electron makes per unit distance traveled in a gas under the influence of an electric field (by applying a voltage):

$$\alpha = \frac{1}{\lambda_i} \text{ and } \alpha = Ape^{-\frac{Bp}{E}} \tag{13}$$

where p is gas pressure, and A and B are constants related to the gas properties.

The created positive ions are also accelerated in the electric field. The positive ions hit the negative electrode surface, generating a secondary emission of electrons. The rate of electrons produced by one ion is the second Townsend's coefficient γ. Secondary electrons can further sustain the ionization processes by contributing to avalanche multiplication. The relation between the first and second Townsend's coefficients is (Braithwaite 2000):

$$\alpha d = \ln\left(1 + \frac{1}{\gamma}\right) \tag{14}$$

where d represents the distance between the electrodes.

Figure 1.6 shows the impact ionization after applying voltage and electron avalanche formation.

The value of the voltage at which the plasma is formed is the breakdown voltage, V_{b}.

$$V_b = \frac{Bpd}{\ln(Apd) - \ln\left[\ln\left(1 + \gamma^{-1}\right)\right]} \tag{15}$$

Equation 15 is Pashen's law (Fig. 5b in Chap. "What are Solution Plasmas?"). At low (pd) values, few collisions occur between electrons and gas atoms or molecules because the pressure is too low or the gap is too small. The low (pd) condition reduces ionization, and a higher V_{b} is needed to cause the breakdown. Too many collisions occur at high (pd) values, and electrons lose energy before ionizing atoms or molecules. In this condition, V_{b} increases. The breakdown voltage reaches a minimum at a specific intermediate (pd) value. At this point, the conditions for ionization are optimal, i.e., the electrons gain enough energy to ionize atoms or molecules. However, not so many collisions occur that the electrons lose the energy prematurely.

4.2 *Cold and Non-LTE Plasmas*

This section will enumerate typical methods to generate cold and non-LTE plasmas. In direct current (DC) discharge, a DC voltage is applied across a pair of electrodes in a low-pressure gas. The electric field ionizes the gas, creating a plasma. This method is used in simple plasma experiments and certain discharge lamps. Plasma can also be created when RF power is applied to a gas at low pressure using suitable electrodes coupled capacitively or inductively. The oscillating electric field at RF frequencies ionizes the gas and forms a plasma. RF plasmas are commonly used in plasma processing applications and industrial systems such as plasma etching and plasma-enhanced chemical vapor deposition. Microwave radiation can ionize gas and generate plasma. Microwave plasmas are common in plasma processing applications, including material synthesis and surface treatment.

Pulsed glow discharge, often used at atmospheric pressure, is generated using a high voltage (about 20 kV) power supply with a pulse frequency typically of 1–30 kHz. The dielectric barrier discharge is similar to the pulsed glow discharge, with the difference that both electrodes are covered with dielectric. In the corona discharge, the cathode electrode is a wire. The gas pressure can vary over a wide range, including atmospheric pressure.

5 Solution Plasma

SP is an electrical gas discharge at atmospheric pressure in the gas formed from the surrounding liquid. SP is a cold and non-LTE plasma with a much higher electron temperature than gas temperature. Gas means atoms, molecules, excited atoms, molecules, radicals, and ions. Water, aqueous solution, organic compound, organic solution, suspension, liquid nitrogen, supercritical fluids, or ionic liquids can be the solution in the SP. In SP, the current–voltage characteristics correspond to the glow discharge curve. As previously described, the low-pressure glow discharge, RF plasma, corona, and barrier discharges at various pressures are cold and non-LTE plasmas. Unlike these discharges, the “wall” of SP is a liquid, i.e., the solution.

The plasma produced in an electric discharge in water has been studied since 1899 to interpret the spectra emitted by stars in astronomy (Wilsing 1899). In 1953, another experiment showed that a discharge in a mixture of methane, ammonia, water, and hydrogen produced many amino acids, hydroxyl, and aliphatic acids (Miller 1953, 1955). Recently, this experiment was reconsidered to demonstrate abiotic amino acid synthesis and focused on explaining the origin and evolution of life on Earth (Jiang et al. 2014). Arc discharges in water have been developed for water treatment and electrical discharge machining (Locke et al. 2006; Jameson 2001).

The SP experiments started in 2008 in the group of Prof. Takai and Prof. Saito at Nagoya University, Japan (Takai 2008).

In a diagram of electron temperature and number density of charged particles, SP fits the glow discharges region with an electron energy of 1–10 eV and an electron density of about 10^{15}–10^{17} m^{-3} (Fig. 3).

Plasma parameters of SP, such as temperature, density, and electric field strength, show specific characteristics due to the surrounding liquid medium, often resulting in behaviors that differ from those of gaseous plasmas. In what follows, we will briefly explain SP parameters.

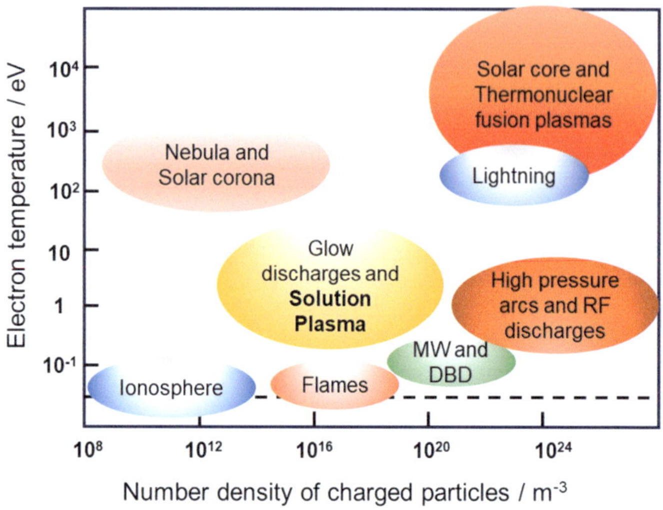

Fig. 3 Diagram of electron temperature and number density of charged particles for different plasmas. Solution plasma fits the glow discharges region with an electron energy of 1–10 eV and a charge density of about 10^{15}–10^{17} m^{-3}

5.1 *Temperature*

In SP, the electron temperature is about 10^4–10^5 K, and the gas temperature is about 300–600 K. Unlike gaseous plasmas, SP experiences significant thermal effects due to the higher heat capacity of the liquid medium. Increased plasma gas temperatures can rapidly heat the liquid, potentially leading to localized boiling or cavitation near the plasma-liquid interface.

In the limits mentioned above, higher temperatures increase the kinetic energy of particles, which enhances reaction rates and drives processes like oxidation, reduction, and nanoparticle formation. However, if temperatures get too high, they may damage sensitive molecules in specific applications like biomedical treatments.

Temperature gradients around the plasma region in SP can lead to complex flows and diffusion effects. The temperature gradients influence the movement of ions, radicals, and reactive species, impacting the uniformity of liquid and the rate of chemical reactions.

5.2 *Electron Density*

In SP, the electron density is about 10^{15}–10^{17} m^{-3}. An increased electron density in SP results in more frequent collisions, which generates more reactive radicals. For example, increased electron density produces more hydroxyl radicals and hydrogen peroxide in water discharge, which are crucial for water purification and disinfection.

Electron density supports ionization rates, producing ions and electrons that enhance chemical reactivity. However, high electron density can lead to more recombination events, reducing the availability of reactive species. In nanoparticle synthesis, electron density affects the rate of reduction and nucleation processes, which determine particle size, morphology, and stability in the solution.

5.3 Electric Field Strength

In SP, the electric field strength directly influences plasma initiation by creating a high voltage to ionize the liquid medium or its vapors. Strong electric fields can create filamentary discharges within the liquid, which are essential in processing applications.

Electric fields accelerate charged particles, creating radicals capable of initiating redox reactions, breaking down organic contaminants, or producing nanoparticles. Electric fields affect the mobility of ions in solution, impacting how charged species interact and transport within the solution. The charge particle movement is critical in electrochemical processes, such as electrolysis, which may operate in tandem with SP.

The electric field strength depends on the electrical conductivity of the solution. The dissolved salts or ions in the solution increase the conductivity of the liquid, which can influence the discharge stability and uniformity. A high conductivity can also reduce the required voltage for plasma generation by providing more charge carriers.

5.4 Formation and Diagnostics of SP

Temperature, electron density, electric field, and composition influence plasma behavior and interactions with liquid mediums in SP. These parameters determine the production of reactive species, energy transfer mechanisms, and the types of chemical reactions possible, enabling various SP applications for nanoparticle and hybrid materials synthesis and water purification.

Chapter "Characteristics of Solution Plasmas" describes in detail the formation and the diagnostics of SP. Section 1 in Chap. "Characteristics of Solution Plasmas" describes the SP formation, which means, first of all, the generation of bubbles in the solution. The next step is the breakdown of the gas from these bubbles. It also explains how the SP can be formed as a streamer discharge depending on the injected power. Section 2 in Chap. "Characteristics of Solution Plasmas" presents the diagnostics of SP. The diagnostics of SP consists of specific methods to measure plasma parameters such as T_e, T_i, n_e, n_i, and n_g, where n_g means atoms or molecules, and radicals density in the SP gas and liquid phases.

References

Becker KH, Kogelschatz U, Schoenbach KH, Barker RJ (2005) Non-equilibrium air plasma at atmospheric pressure. Series in plasma physics. IoP Publishing, Bristol and Philadelphia, Chapter 2, p 23
Bogaerts A, Neyts E, Gijbels R, van der Mullen J (2002) Gas discharge plasmas and their applications. Spectrochim Acta, Part B 57(4):609–658
Braithwaite NSJ (2000) Introduction to gas discharges. Plasma Sources Sci Technol 9(4):517–527
Jiang L, Dziedzic P, Spacil Z, Zhao G-L, Nilsson L, Ilag LL, Córdova A (2014) Abiotic synthesis of amino acids and self-crystallization under prebiotic conditions. Sci Rep 4:6769
Locke BR, Sato M, Sunka P, Hoffmann MR, Chang J-S (2006) Electrohydraulic discharge and nonthermal plasma for water treatment. Ind Eng Chem Res 45(3):882–905
Jameson EC (2001) Electrical discharge machining. Society of Manufacturing Engineers, Michigan
Miller SL (1953) A production of amino acids under possible primitive Earth conditions. Science 117(3046):528–529
Miller SL (1955) Production of some organic compounds under possible primitive Earth conditions. J Am Chem Soc 77(9):2351–2361
Takai O (2008) Solution plasma processing (SPP). Pure Appl Chem 80(9):2003–2011
Wilsing J (1899) On the interpretation of the typical spectrum of the new stars. Astrophys J 10(2):113–125

Characteristics of Solution Plasmas

Nagahiro Saito and Maria Antoaneta Bratescu

Abstract Solution Plasma (SP) is an electrical discharge at atmospheric pressure in the gas formed from a surrounding liquid. Water, aqueous solution, organic compound, organic solution, suspension, liquid nitrogen, supercritical fluids, or ionic liquids can be the solution in the SP. The SP is a cold and non—Local Thermodynamic Equilibrium plasma where the electron temperature is much greater than the gas temperature. The SP operates in a glow discharge regime. The processes inside SP are named Solution Plasma Processes (SPP). SPP combines the physics of atmospheric pressure discharges, the thermodynamic properties of liquids, and the physical and chemical reactions at the interfaces: plasma gas-phase and solution, plasma gas-phase and electrodes surface, and solution-electrodes interface. Four species of particles exist in a plasma gas phase: electrons, ions, neutrals, and radicals. The species densities and temperatures characterize the plasma. This chapter is focused on the formation of the SP and the diagnostics methods using electrical and optical procedures. The electrical characteristics of the SP mean the measurement of the current, voltage, and electron temperature. The optical diagnostics methods like emission and absorption spectroscopy measure the electron and particle densities in the SP gas and liquid phases.

1 Formation of Solution Plasma

Plasmas are ionized gases containing positive and negative charges in equal density and a high density of neutral particles (atoms, molecules, radicals). Plasmas can have various ionization degrees depending on charges and neutral densities. Plasmas can be found in the Universe or can be produced in a laboratory. Depending on the temperature value, the plasmas can be low-temperature plasmas (cold), known as gas

N. Saito (✉)
Department of Chemical Systems Engineering, Graduate School of Engineering, Institute for Innovation for Future Society, Nagoya University, Nagoya, Japan
e-mail: saito.nagahiro.z7@f.mail.nagoya-u.ac.jp

M. A. Bratescu
Institute for Innovation for Future Society, Nagoya University, Nagoya, Japan

N. Saito (ed.), *Solution Plasma Chemistry*, Lecture Notes in Chemistry 115,
https://doi.org/10.1007/978-981-95-0966-9_3

discharges, or high-temperature plasma, as fusion plasmas. Another classification of the plasmas is made according to the temperature equilibrium among the species: electrons, ions, and neutrals. If all the species from plasma have the same temperature, the plasma is Local Thermodynamic Equilibrium (LTE) plasma. Otherwise, it is non-LTE, i.e., different plasma species have different temperatures. In a non-LTE plasma, the electrons have a much higher temperature than the gas. Gas means atoms, molecules, ions, and radicals.

Other plasma classifications can be done depending on experimental conditions: the chemical input as the working gas, the pressure which can vary from 0.1 Pa to atmospheric pressure, the electromagnetic field used for plasma ignition and operation, including the temporal behavior, and the discharge configuration.

Solution Plasma (SP) is an electrical gas discharge at atmospheric pressure in the gas formed from the surrounding liquid. Water, aqueous solution, organic compound, organic solution, suspension, liquid nitrogen, supercritical fluids, or ionic liquids can be the solution in the SP. The SP is a cold and non-LTE plasma and operates in a glow discharge regime. The SP name comes from various plasmas obtained by choosing different combinations of solvents and solutes (Takai 2008, Takai et al. 2013, Saito et al. 2018, and Saito et al. 2019). The processes inside SP are named Solution Plasma Processes (SPP). SPP combines the physics of atmospheric pressure discharges, the thermodynamic properties of liquids, and the physical and chemical reactions at the various interfaces: plasma gas and solution, plasma gas and electrode surface, and solution and electrode surface.

1.1 Solution Plasma as a Non-local Thermodynamic Equilibrium Plasma

As mentioned in Sect. 2.5, the plasma produced in an electric discharge in water has been studied since 1899 by Wilsing to interpret the spectra emitted by stars. Later, a discharge in a mixture of methane, ammonia, water, and hydrogen produced many amino acids, hydroxyl, and aliphatic acids (Miller 1953 and Miller 1955). The Miller experiment was recently repeated to demonstrate abiotic amino acid synthesis and explain the origin and evolution of life on Earth (Jiang et al. 2014). One of the most developed applications of plasma in liquids was the arc discharge in water used for water treatment and electrical discharge machining (Locke et al. 2006 and Jameson 2001).

The SP is a cold and non-LTE plasma with a much higher electron temperature (T_e) than gas temperature (T_g). The SP current–voltage characteristics correspond to the glow discharge curve. The low-pressure glow discharge, RF plasma, corona, barrier, and hollow cathode discharges at various pressures are non-LTE plasma. In non-LTE plasmas, T_e can reach the value of 20,000 K, while T_g is slightly greater than room temperature.

Since collisions cause a continual exchange of energy between electrons and heavy particles with a tendency to equilibrate temperatures, it is challenging to maintain non-LTE conditions at elevated pressure with high collision rates and short mean free paths. For steady-state plasmas, the deviation from LTE can be expressed by the following formula $\frac{T_e - T_g}{T_e} = \frac{m_g}{4m_e}\left(\frac{\lambda_e eE}{\frac{3}{2}k_B T_e}\right)^2$, already explained in Chapter "Basics of Plasma Physics", Eq. 4.

Figure 1 displays the simple arrangement of the SP used in the laboratory. A bipolar pulsed power supply is connected to a pair of electrodes immersed in the solution. The standard conditions are shown in the figure. Usually, tungsten is used for electrodes with an electrode gap distance of 0.5 mm. In other experiments, electrode erosion is needed, and other metal materials are used as electrodes. The repetition frequency is set from 15 to 200 kHz. The pulse width is typically 1–3 μs. Under this condition, the solution plasma can be generated between electrodes. The SP setup is a straightforward experiment that offers many challenges in material science.

As mentioned, the SPP combines the physics of atmospheric pressure discharges, the thermodynamic properties of liquids, and the physical and chemical reactions at the interfaces: plasma gas-phase and solution, plasma gas-phase and electrodes surface, and solution-electrodes interface. The electron avalanche in the discharge in a solution is due to a phase change, i.e., the liquid turns to the gas phase, where the avalanche of electrons occurs. In liquid solutions, plasma processes are more complex than those produced in the gases surrounded by a solid wall. The physical and chemical processes contribute to the discharge in the gas phase, the liquid phase, the plasma itself, and the interfaces between plasma and liquid, electrode-liquid, and electrode-plasma (Saito et al. 2018, 2019). The most important phenomena are the bulk processes in the liquid medium and the surface processes at the electrode-liquid and plasma-liquid interfaces (Ceccato et al. 2010; Lewis 1994).

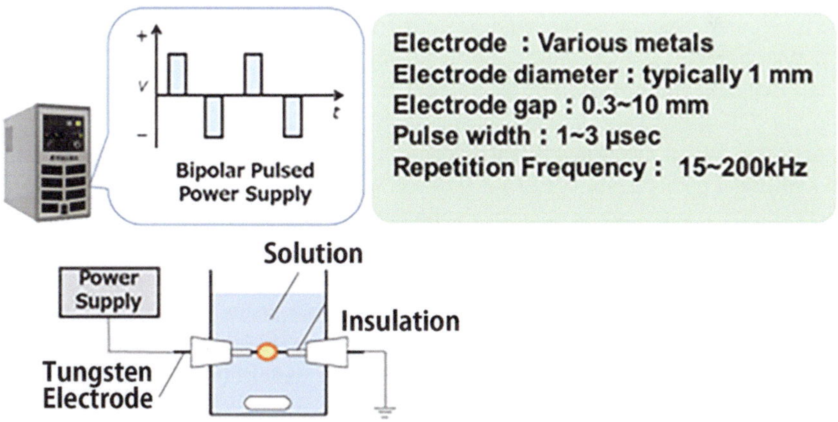

Fig. 1 A simple arrangement of solution plasma was used in the laboratory. A bipolar pulsed power supply is connected to a pair of electrodes immersed in the solution. The insert contains the typical conditions. The conditions can vary depending on the purpose of the experiment

1.2 Bubbles Formation

The bubble breakdown theory explains that the electric discharge in solutions occurs in preexisting gas bubbles or bubbles produced by electrolysis near the electrodes, boiling, or decomposition due to electrons. The electronic mechanism of the breakdown shows that the discharge results from the electron multiplication in the avalanche in solution, similar to the generalized gas discharge pattern. In the liquid phase, the electrons are accelerated in high-intensity fields, producing ionization of molecules and atoms. The electronic mechanism better describes dielectric liquids breakdown than conductive solutions.

The gas-phase formation in liquid, the appearance of a bubble near the electrode tip, and the gas breakdown will be discussed. When an electric field is applied, a gas bubble forms. In the beginning, accelerated electrons are produced, which determine the local heating and vaporization of the liquid. Accelerated electrons have higher energy than thermal energy, making ignition by multiplying electrons. Local heating, cavitation, and electrical stress co-occurring in solutions can explain the bubble formation. The properties of the solution, like conductivity, pH, and temperature, control the mechanism of the bubble formation (Kattan et al. 1989, Murphy 1970, and Kattan et al. 1991).

An avalanche of electrons occurs in this gas bubble, and the amplification processes begin. The electronic avalanche is controlled either by injecting several electrons into the gas phase or by ionization of the molecules from bubbles due to the increased electric field produced in this region due to a decreased dielectric constant.

Lewis showed that the mechanism of bubble formation is a mechanical process. The electric field of order 10^8–10^9 V/m produces a mechanical deformation, which determines the temperature increase in the mechanical deformations in the solution at the electrode-solution interface and the solution. The mechanical deformations in the solution represent the areas that allow the formation of "a crack." The crack in the molecular structure of the solution consists of a low density or even a region of low pressure, called a "vacuole," into which electrons can be injected. Mechanical stress explains the formation of the vacuole in the area where the electric field was applied. The appearance of vacuoles and bubbles inside the solution defines plasma generation. When electrons are introduced into the crack, the crack becomes like an additional cathode, similar to a newly formed micro protrusion. The appearance of a "crack" in the solution determines the ignition process of the discharge (Lewis 1988).

Bubbles ranging from units to tens of microns may also exist in the solution, mainly on electrodes. The ionization processes occur at the beginning inside the bubble due to the electric field when the applied voltage has the value of the ignition (breakdown) voltage V. After the discharge ignites, the electric field in the bubble decreases.

The decrease in voltage causes the attenuation or cessation of ionization processes. The bubble produces electric charges, causing bubble elongation along the field and penetration of the electric charge into the solution at a speed determined by the mobility of the charge carriers. Two situations can be distinguished: sustained or cessation of the discharge.

The voltage applied to the bubble follows Paschen's law in the first case (Paschen's law shown in Fig. 5 Chapter "What are Solution Plasmas?" will be discussed for SP in Sect. 1.3). In the second case, the voltage applied to the bubble increases, causing the discharge to end (Ushakov et al. 2007a, b).

In dielectric solutions, particles such as suspensions may exist as impurities. The particle has electric charges and polarizability in an electric field. Under these conditions, the particle will deviate in the electric field. The particle deviation and velocity depend on the electric field and the solution properties (viscosity and dielectric constant). When it reaches the electrode, the particle changes the local electric field, causing a change in the structure of the solution.

1.3 Breakdown in Bubbles; Townsend Discharge and Paschen's Law

In gas plasma, the physical processes of discharge have been studied well. At low values of *pd* (the pressure multiplies by the electrode gap distance), the breakdown is due to the avalanche of electrons, as explained by Townsend. At high values of *pd*, the streamer mechanism explains the discharge ignition (Schönland 1956). In liquids, the high density of such medium hinders electrons from accelerating and producing collisional processes, such as excitation, dissociation, and ionization, unless the electric field is a few orders of magnitude larger than that of plasma at atmospheric pressure (Ushakov et al. 2007a, b).

The electron avalanche is supposed to hardly occur inside a liquid due to the high recombination rate and reduced mobility of charges. An explanation is needed regarding the breakdown in solutions and plasma operation. In solutions, a phase change occurs, i.e., the liquid is transformed locally into a gas, and then in the gas, the avalanche of electrons is produced. Thus, the thermodynamics of liquids and gases play a significant role, and the solution plasma can be connected with the physics of gas discharges.

The breakdown in gas plasma starts with understanding the first ionization coefficient. A simple arrangement consists of two parallel electrodes at a distance d from where an insulating gas exists (Fig. 2). When no electric field is applied between the electrodes, a balance between the electrons and the positive ions exists. The steady state is disturbed when an external electric field is applied between electrodes. Townsend studied the current (I) as a function of the applied voltage (V) (Fig. 3). At low values of the applied voltage, the electric current increases proportionally with the rise of the voltage and then remains constant at the value of the saturation current I_0. As the voltage increases, the electric current increases exponentially. The exponential growth of the current is due to the gas ionization by electron collisions with atoms or molecules from the gas. The electric field, i.e., V/d ratio, increases by increasing the voltage. As a result, the kinetic energy of the electrons increases, increasing the number of collisions that produce ionization (Townsend 1910 and Wadhwa 2020).

Townsend explained the exponential increase in current and introduced the coefficient α to represent the number of electrons produced per electron per unit of length along the electric field. Let us denote n_0 as the number of electrons leaving the cathode. At the distance x from the cathode, the number of electrons is n. In the interval dx, the electrons produce additional electrons by ionization collisions. Therefore,

$$dn = \alpha n dx$$

To solve this differential equation, we first separate the variables n and x by dividing both sides of the equation by n:

$$\frac{dn}{n} = \alpha dx$$

Next, we integrate both sides:

$$\int \frac{1}{n} dn = \int \alpha dx$$

This yields the natural logarithm on the left side and a linear term on the right side:

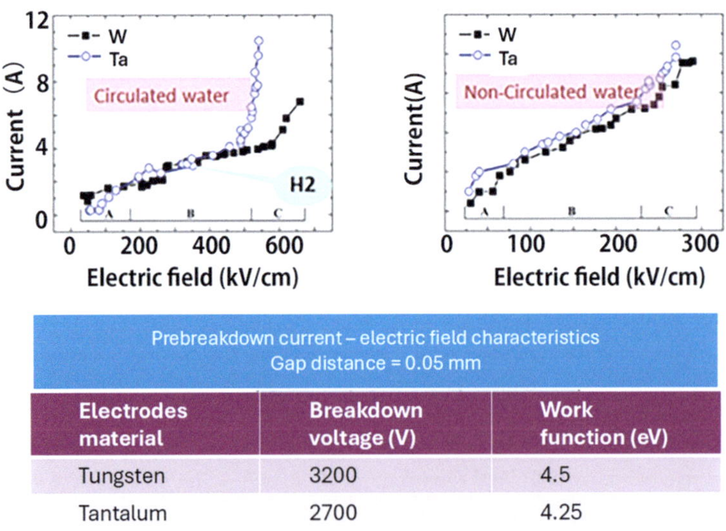

Fig. 2 A parallel plate capacitor filled with an insulating gas separated by a distance d

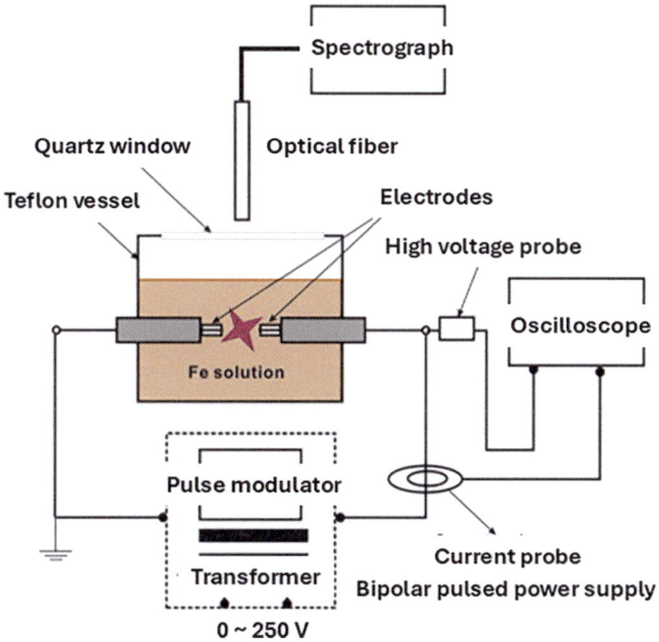

Fig. 3 Dependence of the current as a function of the applied voltage between the electrodes

$$\ln n = \alpha x + \mathrm{A}$$

where A is the constant of integration. Now, using the boundary condition that at $x = 0$, $n = n_0$, we substitute these values into the equation to determine A:

$$\ln n = \alpha x + A$$

Thus, the equation becomes:

$$\ln n = \alpha x + \ln n_0$$

or equivalently, by exponentiating both sides to remove the logarithms:

$$n = n_0 e^{\alpha x}$$

At $x = d$, the number of electrons is:

$$n = n_0 e^{\alpha d}$$

This means the number of electrons increases exponentially as they travel from the cathode to the anode. Finally, the current I is proportional to the number of electrons, so the current at $x = d$ is given by:

$$I = I_0 e^{\alpha d} \tag{1}$$

The term $e^{\alpha d}$ represents the electron avalanche and describes the exponential multiplication of electrons as they move from the cathode to the anode (Townsend, 1910).

The cathode plays an essential role in plasma by producing electrons to ignite and maintain the discharge. The second ionization coefficient describes the processes. Electrons cannot leave the crystalline metal structure due to the electrostatic forces between ions and electrons. The energy required for an electron to leave the metal is the work function determined by the Fermi level characteristic of each metal. There are different ways in which energy can be provided to release the electron. Thermionic and field emissions are not processes suitable for the solution plasma. However, electron emission by positive ions and excited atom bombardment is a process in plasma ignition in the solution plasma.

The following section will discuss the electron emission by the bombardment with positive ions and excited atoms. The electrons can be emitted by bombarding the cathode surface with accelerated positive ions in the electric field applied between the electrodes. This process is known as *secondary emission*. During the secondary emission, the energy of the positive ion W must be twice the work function (minimum thermodynamic energy to remove an electron) of the metal because the electron will neutralize the bombarded positive ion, and the other electron will be released. The secondary electron emission occurs if $W_\mathrm{k} + W_\mathrm{p} \geq 2W$, where W_k and W_p are the positive ion kinetic and potential energy.

The Townsend spark discharge mechanism consists of the emission of electrons by the positive ion. Electrons can also be released from the cathode surface by interacting with excited atoms or molecules at metastable levels. These processes are more complicated depending on the nature of the atoms, molecules, and cathode material (Townsend 1910).

Townsend observed that the plasma current produced between two parallel electrodes increases much faster than the value given by Eq. 1. when the applied voltage increases (Fig. 4). To explain this deviation, Townsend suggested that the extra current must be due to the presence of positive ions and photons. Positive ions release electrons by collisions with gas atoms or molecules and bombarding the cathode. Similarly, photons produce electrons after collisions with atoms or molecules in the gas and cathode due to the interaction between the photon and the cathode material. The phenomenon of self-sustaining discharge in which electrons are released from the cathode by bombardment with positive ions must be considered.

Let n_0 be the number of electrons produced from the cathode due to ultraviolet radiation, n_+ the number of electrons produced from the cathode due to the bombardment with positive ions, and n the number of electrons that reach the anode. Let γ be defined as the number of electrons released from the cathode per incident positive ion. The coefficient γ is the second Townsend ionization coefficient. In this case

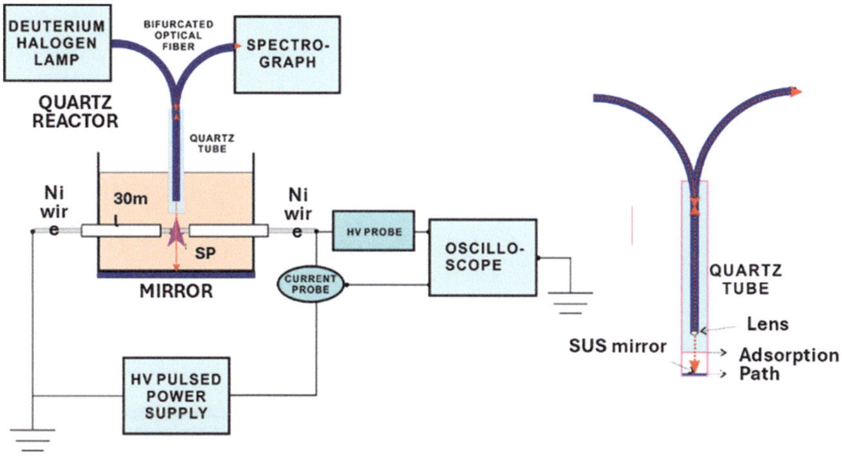

Fig. 4 Variation of gap current for different electrode spacing in uniform E. From Eq. 1, $\ln I = \ln I_0 + \alpha x$. This relation is the straight line with the slope α and intercepts $\ln I_0$

$$n = (n_0 + n_+)e^{\alpha d}$$

The number of electrons in the gas is $n-(n_0 + n_+)$. Corresponding to each electron produced in the gas, there will be one positive ion. Assuming that each positive ion produces γ effective electrons from the cathode, then $n_+ = \gamma\left[n - (n_0 + n_+)\right]$ and $n_+ = \frac{\gamma(n-n_0)}{1+\gamma}$. Substituting n_+ in $n = (n_0 + n_+)e^{\alpha d}$ the following expression is obtained $n = \frac{n_0 e^{\alpha d}}{1-\gamma\left(e^{\alpha d}-1\right)}$.

In terms of current, the expression is

$$I = \frac{I_0 e^{\alpha d}}{1 - \gamma\left(e^{\alpha d} - 1\right)} \quad (2)$$

An ion traveling in the electric field between the electrodes gains energy. A pair of ions can be produced in the collision process between an atom or molecule with an ion. In the expression of the discharge current, Townsend introduced the term β, representing the number of ion pairs produced by a positive ion traveling a 1 cm path in the direction of the field.

$$I = I_0 \frac{(\alpha - \beta)e^{(\alpha-\beta)d}}{\alpha - \beta e^{(\alpha-\beta)d}} \quad (3)$$

Because of $\beta \ll \alpha$, Eq. 3 becomes Eq. 2 with $\gamma = \beta/\alpha$.

Townsend, in 1910, and Little, in 1956, explained the breakdown mechanism. When the voltage between electrodes increases, the discharge current is $I = \frac{I_0 e^{\alpha d}}{1-\gamma\left(e^{\alpha d}-1\right)}$. The current becomes infinite when $\gamma\left(e^{\alpha d} - 1\right) = 1$.

Because of $e^{\alpha d} \gg 1$, the discharge current is the same as the current in the whole electrical circuit. The plasma current is limited by the power supply impedance and an external impedance. The condition $\gamma\left(e^{\alpha d}-1\right)=1$ is the Townsend breakdown criterion or the Townsend criterion for spark formation.

The current between the electrodes is not self-sustaining as long as the denominator $1-\gamma\left(e^{\alpha d}-1\right)$ is positive. When the electric field and the Townsend coefficient α become large enough, the denominator $1-\gamma\left(e^{\alpha d}-1\right)$ goes to zero, indicating the transition to self-sustaining discharge, and thus the breakdown occurs. The most straightforward condition for the gas breakdown between the electrodes under an electric field is

$$\gamma\left(e^{\alpha d}-1\right)=1 \text{ or } \alpha d=\ln\left(\frac{1}{\gamma}+1\right) \tag{4}$$

Townsend gives a semiempirical formula for the α coefficient:

$$\frac{\alpha}{p}=A\ \exp\left(-\frac{B}{E/p}\right) \tag{5}$$

where p is pressure, E is the electric field between the electrodes, and A and B are constants that can be experimentally determined (see Eq. 13 in Chapter "Basics of Plasma Physics"). From Eqs. 4 and 5, the breakdown voltage (V_b) as a function of the similarity parameter pd is calculated to be

$$V_b=\frac{B(pd)}{\ln(Apd/K)} \tag{6}$$

where $K=\ln\left(\frac{1}{\gamma}+1\right)$. The dependence of the breakdown voltage on the product pd is known as Paschen's law. The Paschen curve gives a minimum value of the breakdown voltage $V_{\min}=eV_{\min}=\frac{B}{A}\ln\left(1+\frac{1}{\gamma}\right)$ at $pd_{\min}=e\frac{\ln\left(1+\frac{1}{\gamma}\right)}{A}$, where $e\approx 2.718$ is Euler's number (see Fig. 5b in Chapter "What are Solution Plasmas?" and Eq. 15 in Chapter "Basics of Plasma Physics").

A qualitative explanation of the minimum breakdown voltage considers the ionization efficiency of atoms and molecules by collisions with electrons depending on the electron energy. Consider that the pressure is constant. If the second Townsend ionization coefficient γ is small for the $pd>(pd)_{\min}$, the electrons crossing the distance from the cathode to the anode produce more collisions with the gas atoms and molecules than at $(pd)_{\min}$. However, the energy gained between successive collisions is less than at $(pd)_{\min}$. Therefore, the ionization probability is lower if the voltage does not increase. In the case of $pd<(pd)_{\min}$, the electrons cross the distance from the cathode to the anode almost without ionization collisions, and thus the breakdown voltage is high. Therefore, the point $(pd)_{\min}$ corresponds to the highest ionization efficiency by electron collisions with atoms and molecules, yielding the minimum breakdown voltage (Braithwaite 2000). Figur 5b in Chapter "What are Solution Plasmas?" displays the dependence of breakdown voltage on (*pd*) in different atomic and molecular gases.

1.4 Streamer Discharge

The electric discharge mechanism of liquid dielectrics has been a topic of great interest for scientific and practical reasons. It has been observed that positive streams (discharge channels) occur in nanoseconds to microseconds in the pre-discharge stage. Optical measurement techniques record such ultrafast events. A small pre-discharge current of the order of microamperes to milliamperes exists when high voltage is applied until plasma formation. Plasma formation occurs with a sudden increase in current. The delay between the voltage application and the plasma production is mainly associated with the time required to propagate a streamer between the electrodes (Raizer 1987; Yang et al. 2012).

The Townsend discharge breakdown mechanism is generally successfully applied for low *pd* values. The spark and streamer mechanisms are involved in the case of high *pd* values. In spark and steamer mechanisms, the discharge occurs in a narrow channel where the current density becomes very high. The secondary electron emission from the cathode is low. Therefore, the cathode material does not influence the breakdown.

Both spark and streamer breakdown and the Townsend mechanism breakdown are primarily related to electron avalanches. However, ignitions in large gaps cannot be considered independent and are stimulated by the emission of electrons from the cathode. The breakdown at high values of *pd* and the application of a large electric field occurs much faster than the time required for the ions to cross the distance between the electrodes. That is why the secondary emission of electrons is very weak. The spark and streamer channel with high electrical conductivity can be formed faster than electron travel from cathode to anode (Yang et al. 2012, p 38 and Wadhwa 2020).

The electric charges between the electrodes at a distance d increase by a factor $e^{\alpha d}$ when the field between the electrodes is uniform. The above assumption is valid only if we suppose that the electric field $E_0 = V/d$ is not influenced by the space charge created by electrons and positive ions. If the electric charge has a concentration greater than 10^6 yet less than 10^8 m^{-3}, the production of an electronic avalanche is diminished, i.e., $\frac{dn}{dx} < e^{\alpha d}$. Whenever the concentration of electric charges is higher than 10^8 m^{-3}, the avalanche current is followed by a steep increase in the electric discharge current. Depending on the electron concentration, the intensity of the avalanche is influenced by the modification of the electric field E_0 concerning the space charge created between the electrodes (Fig. 5).

The higher mobility of electrons relative to positive ions causes the space charge to be negative and concentrated in a spherical volume. The electric field E_0 is reduced depending on the space charge. If the space charge is more significant than 10^6 m^{-3}, the change in the electric field is substantial. A change in the electric field of 1%

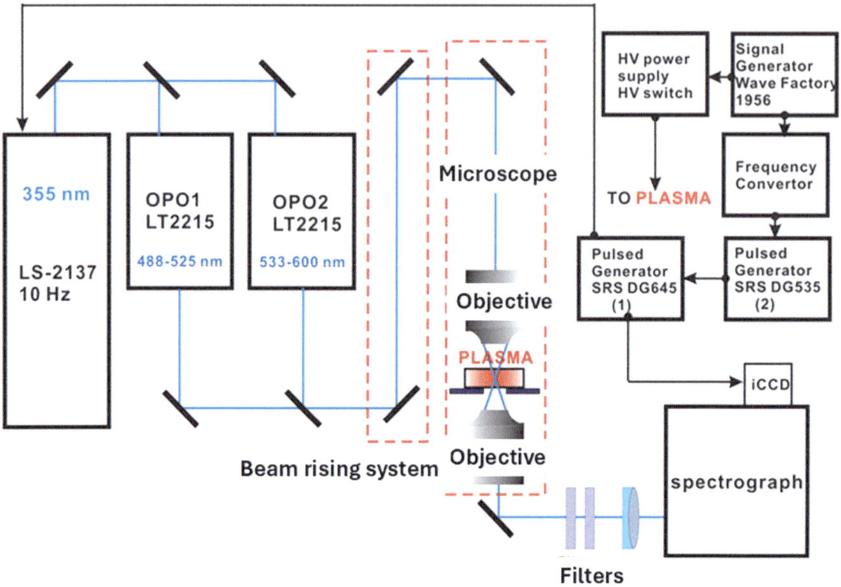

Fig. 5 The redistribution of the electric field due to space charge

causes a doubling of the avalanche. If the space charge exceeds 10^8 m^{-3}, the electric field of the space charge is close to $E_{0,}$ and the streamer occurs. Therefore, the spatial load density plays an essential role in the formation of the streamer.

The external electric field distortion due to the space charge of the dipole is shown in Fig. 6. In front of the avalanche head, the external E_0 and internal E' electric fields make a total field stronger, accelerating ionization. Conversely, the entire electric field is reduced between the separated charges, or "inside the avalanche," slowing the ionization. At some critical value of αx, the electric field induced by charge $N_e \approx \exp(\alpha x)$ reaches the value of the external field E_0. Then, the avalanche-to-streamer transformation occurs when the internal field of an avalanche becomes comparable with the external electric field (e.g., when the amplification parameter αx is sufficiently large).

Theoretical work has been done to use the Townsend equation for the spark discharge breakdown in the air at pressures close to the atmosphere. The mechanisms of the discharge ignition in the air at high pressures, i.e., the electron avalanche, gas ionization, and positive streamer, were demonstrated. A quantitative criterion for streamer formation was applied to give the theory for the breakdown of sparks in the air into atmospheric pressure. The theory agrees quantitatively with the experiment and predicts deviations from Paschen's law. The theory applied to longer sparks at atmospheric pressures, and the effect of decreasing photoionization in longer avalanches was considered. The approach developed in these references might be used to explain SP breakdown (Yang et al. 2012, p 39). Loeb's papers considered the mechanism of spark discharge in air at atmospheric pressure (Loeb and Meek

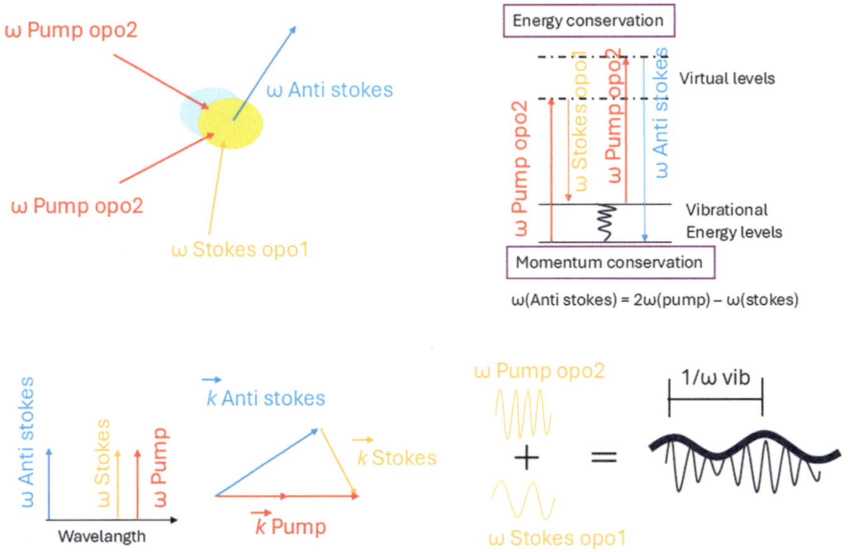

Fig. 6 Electric field distribution in the avalanche. The external electric field distortion is due to the space charge of the dipole

1940a, b). These theories might be well applied to the SP, a plasma at atmospheric pressure.

1.5 Breakdown in Solution Plasma

Many experiments investigate and characterize the SP generated in water and water solutions. Two distinguished stable modes of the discharges generated in the water solutions can be obtained depending on the inter-electrode distance: connected (bridged) and separated discharges (unbridged). The water conductivity, pulse frequency, and initial water temperature were measured in bridged and unbridged operation modes. Optical Emission Spectroscopy (OES) was used to study the SP and to correlate the data between the SP species in the gas phase with the chemical species measured in the solutions under various parameters.

Baroch et al. 2011 paper studied the dependence of the breakdown voltage on various SP parameters (Fig. 7). The high number of charge carriers with increasing conductivity leads to higher current in pre-breakdown and consequently to the ignition of the discharge at a lower applied voltage. The breakdown voltage decreases if the pulse repetition frequency increases due to higher energy injected in the SP, as shown in Fig. 7b. In the case of electrode distance being 16 mm, the breakdown voltage is slightly higher than in the case of interelectrode-distance of 4 mm because the ohmic water resistance between electrodes for a distance of 16 mm is higher than in the case of 4 mm. Because of the increased temperature, the conductivity also increases. The experiment ensured the same conductivity in all measurements to get only the effect of temperature. The procedure is described in the following. First, the required temperature was set, and then the conductivity was adjusted by KCl to 1000 μS cm^{-1}. Figure 7c shows a slight influence of the temperature on the breakdown voltage. The internal heating of the space between the electrodes caused by the flowing current (Joule effect) is significantly higher than the heat delivered from the external source. Therefore, no more significant effect of the water temperature on the breakdown voltage was observed. However, it must be noted that this is valid only if the solution has the same conductivity as the initial conductivity of 1000 μS cm^{-1} at a temperature of 20 °C.

A constant interest is improving the SP operation for better application in new material synthesis. A low-pass filter circuit was inserted into the power supply to increase the density and excitation temperature of the electrons involved in the SP non-equilibrium plasma. The properties of the SP generated by the conventional and the low-pass filter circuits were investigated. The filtration of the MHz frequency range current oscillations increases the maximum discharge current and offers higher stability to the SP. The filter consists of a resistor–capacitor (RC) component inserted

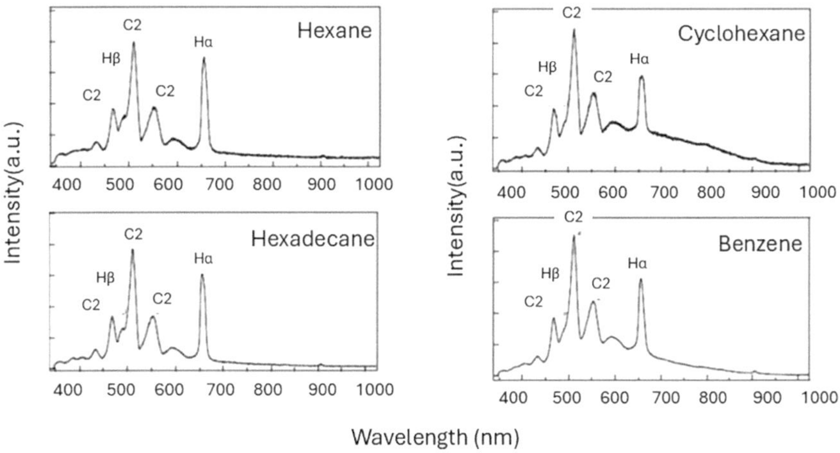

Fig. 7 **a** The breakdown voltage as a function of the water conductivity σ for separated (unbridged) and connected (bridged) discharge modes. The experimental conditions are the pulse repetition frequency of 10 kHz, the pulse width of 2 μs, the initial solution temperature of 20 °C; **b** The breakdown voltage as a function of the repetition frequency for separated (unbridged) discharge mode. The experimental conditions are the pulse repetition frequency of 2 μs, the solution conductivity of 1000 μS cm^{-1}, the initial solution temperature of 20 °C; **c** The breakdown voltage for both separated (unbridged) and connected (bridged) discharge modes as a function of the water temperature. The experimental conditions are the pulse repetition frequency of 10 kHz, the pulse width of 2 μs, and the solution conductivity of 1000 μS cm^{-1} (reprinted from Baroch et al. 2011 Copyright, with permission from IOP publishing)

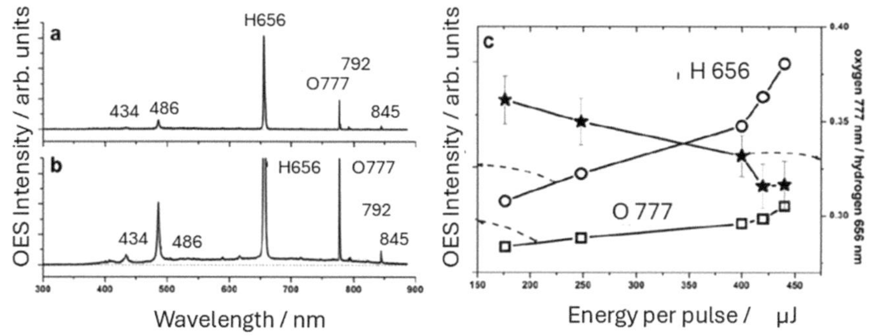

Fig. 8 The solution plasma setup operates with **a** the conventional circuit and **b** the low pass filter circuit, (reprinted from Heo et al. 2017 Copyright, with permission from Wiley Publishing)

in the circuit, which leads to an increase in the density and the temperature of electrons (Heo et al. 2017) (Fig. 8).

Different stages before and after the breakdown of solution plasma were observed (Fig. 9):

(i). The initial condition: The voltage is 0 V, and the current is 0 A.
(ii). The voltage is 1600 V for a 2 µs pulse width, and the current is a few microamperes.
(iii). The voltage steeply decreases to a low value due to the breakdown, and the current abruptly increases.
(iv). The plasma is generated. The discharge current flows with a maximum value of 6–13 A, depending on the external circuit.

The optical emission spectroscopy (OES) data show that the modified solution plasma circuit increases the electron energy and effectively synthesizes materials with high-quality functions.

During the SP breakdown, the bubble formation plays an important role (Baroch et al. 2008). An experiment in the SP explains that injecting bubbles into the solution considerably decreased the breakdown voltages. Two systems, based on a wire-to-wire electrode configuration, (i) with and (ii) without gas bubbling, were designed for investigation. Both systems were operated utilizing a 20 kHz bipolar pulse power supply. The electrical and OES measurements examined the properties of the generated SP. The most important results are displayed in Fig. 10. The injection of the bubbles determines a decrease in the breakdown voltage. In the formation process of SP, the energy of the power supply is used to produce bubbles and plasma breakdown. If bubbles are provided, smaller voltage and less energy are necessary from the power supply for the SP breakdown. Figure 10 shows the current–voltage characteristics of the SP with and without air bubbles as a function of the solution conductivity.

A paper from 2010 explains the influence of the electrode work function on the SP breakdown and operation. The breakdown of the SP initiated in ultrapure water between pairs of tungsten and tantalum electrodes was investigated in two situations

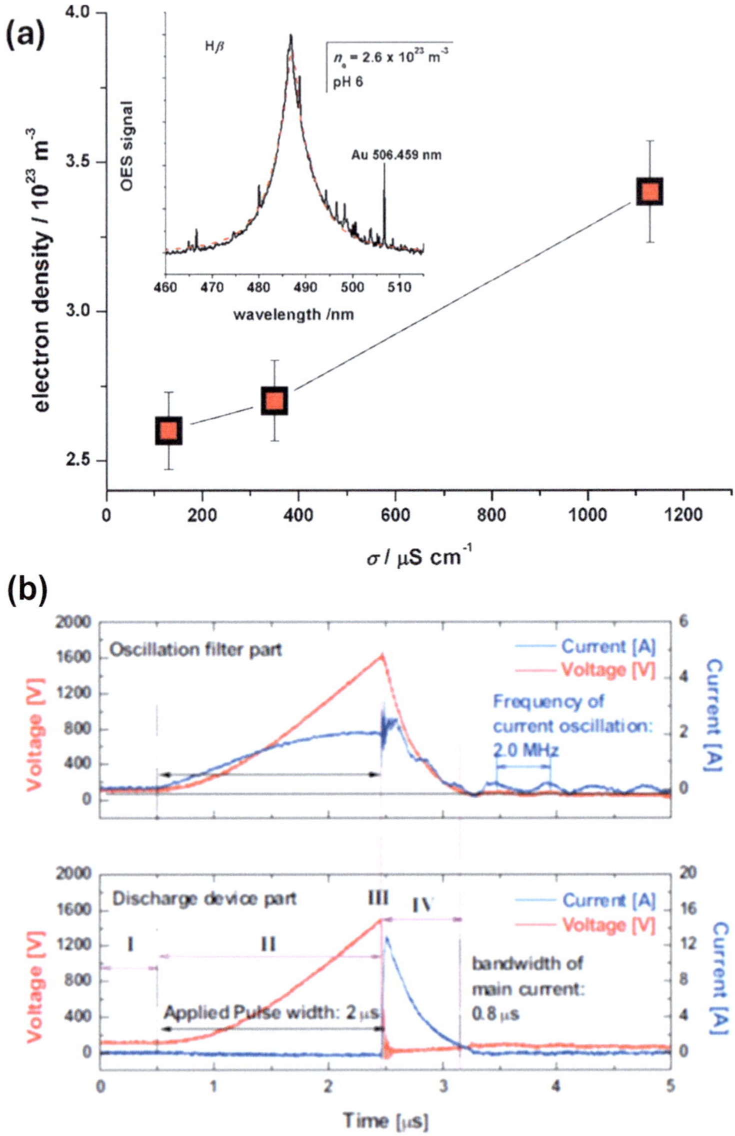

Fig. 9 Current–voltage waveforms of the solution plasma discharge with **a** the conventional circuit and **b** the low-pass filter circuit. The top and bottom waveforms were obtained from probes (1) and (2) shown in Fig. 8b, respectively (reprinted from Heo et al. 2017 Copyright, with permission from Wiley Publishing)

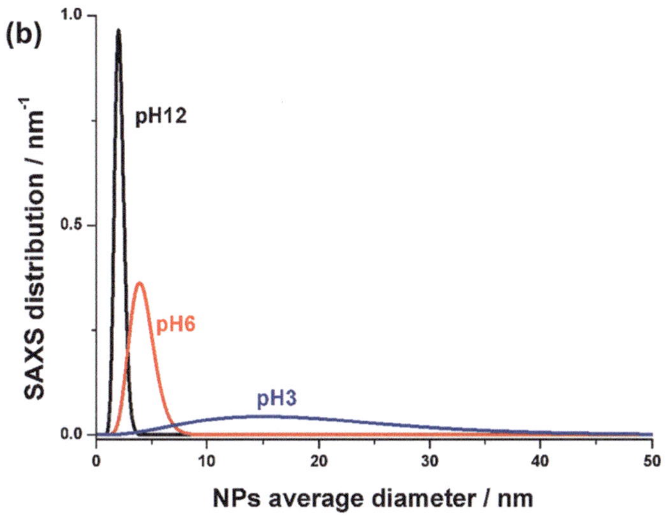

Fig. 10 Current–voltage characteristics of the SP with and without air bubbling as a function of the solution conductivity. The experiments were performed using a bipolar pulsed power supply at a frequency of 5 kHz; the distance between the electrodes was 0.25 mm, (reprinted from Baroch et al. 2008 Copyright, with permission from Elsevier Publishing)

when water was circulated and not circulated. The current–voltage characteristics presented the aspect of a spark discharge for both types of electrodes used in the process. Figure 11 represents the prebreakdown current–electric field characteristics for (a) circulated water and (b) not-circulated water in a 0.05 mm electrode gap distance in the case of W and Ta electrodes (Miron et al. 2010).

Three distinct regions (A, B, and C) could be observed in the current characteristics of the dependence on the electric field in both experimental conditions (Fig. 11). The maximum current values, the bubbles formation, and the values of the water pH characterize these regions. Spikes of current or random pulses, accompanied by light spots, could not be seen in region A but were observed in region B (Fig. 11) and more frequently when Ta electrodes were used in the SP. In region B, bubbles formed between electrodes. The electric current became high in region C, and the size of the bubbles increased.

After forming the bubbles, the avalanche of electrons occurs, leading to the water molecule ionization and dissociation. The electric field required for discharges was about 600 kV/cm for the circulating water and about 300 kV/cm for the uncirculated water.

Figure 12 shows the water pH measured in time in the not-circulated water in the prebreakdown when the applied voltage was gradually increased. The water pH at the beginning of the experiments was 7.05. In region A, the pH decreased, even at low values of the applied electric field. Therefore, the ionization and dissociation of the water molecules occurred before the plasma breakdown, similar to electrolysis: $H_2O + e^-_{fast} \rightarrow H^+ + OH^- + e^-_{slow}$.

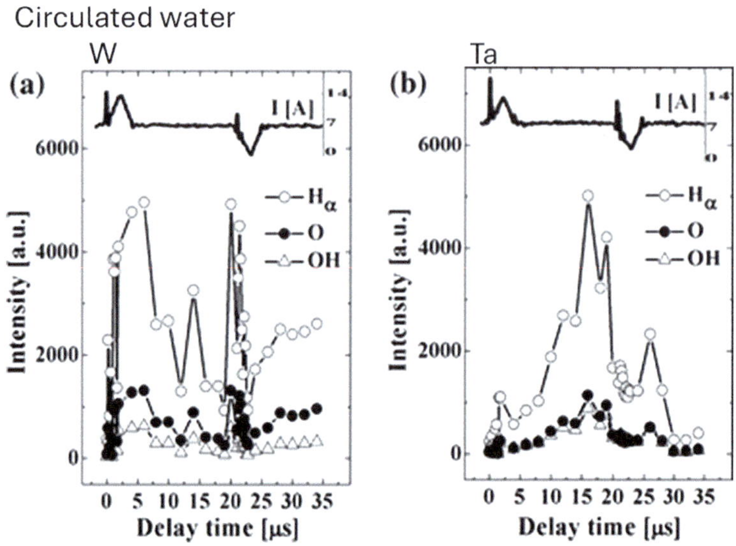

Prebreakdown current – electric field characteristics
gap distance = 0.05 mm

Electrodes material	Breakdown voltage [V]	Work Function (eV)
Tungsten	3200	4.5
Tantalum	2700	4.25

Fig. 11 Prebreakdown current–electric field characteristics for circulated water (**a**) and not-circulated water (**b**) in a 0.05 mm gap between W and Ta electrodes (reprinted from Miron et al. 2010 Copyright, with permission from Springer Publishing)

In region B, the water pH continued to decrease with the increase of the applied voltage (Fig. 12). The water conductivity measured in this region was about 3 μS/cm for the W electrodes and about 2 μS/cm for the Ta electrodes. After that, the current density increased, and current spikes appeared between the electrodes. The local temperature also increased in the interelectrode gap, determining the formation of the bubbles. The bubble generation continued in region C, the ionization of the gas, and the avalanche process leading to the plasma breakdown. The water pH values measured for W electrodes were lower than for Ta electrodes. According to the Pourbaix potential-pH equilibrium diagrams, Ta is more easily oxidized than tungsten at a solution pH of 7. The oxide layer acts as a dielectric layer, so more energy is required for electron extraction from the Fermi level, which is the case for the Ta electrodes. Therefore, the amount of ions produced in the water is lower than in the case of the W electrodes. The formation of a dielectric layer on Ta electrodes explains the higher voltage required for the plasma breakdown than the values measured in the case of W electrodes (Fig. 11).

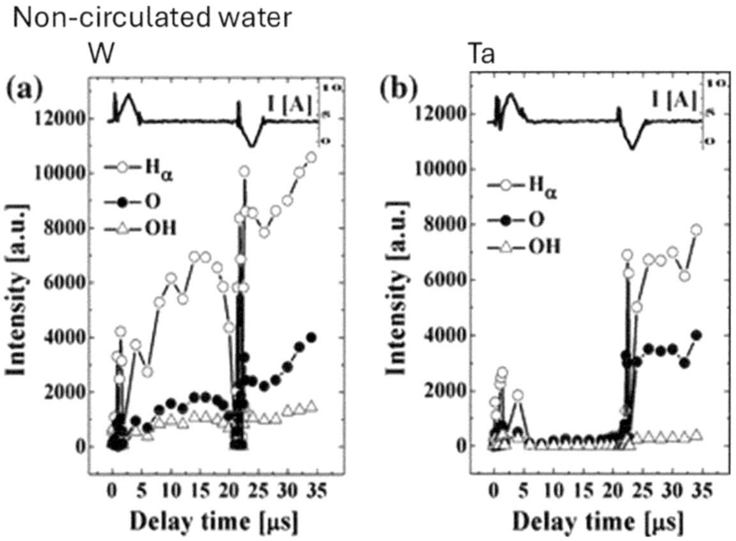

Fig. 12 **a** and **c** water pH values measured in the prebreakdown and after breakdown, respectively, in the not-circulated system when an electric field was applied to the W and Ta electrodes before breakdown; **b** water conductivity after the breakdown. (reprinted from Miron et al. 2010 Copyright, with permission from Springer Publishing)

2 Solution Plasma Diagnostics

Four species of particles exist in a plasma gas phase: electrons, ions, neutrals, and radicals. A radical is any molecular or atomic particle capable of independent existence that contains unpaired electrons in the molecular orbital. The presence of unpaired electrons results in specific properties as unstable and highly reactive behavior. The species densities and temperatures characterize the plasma. This subchapter will discuss the electrical and optical methods used to diagnose solution plasma. The electrical tools enable the measurement of the current–voltage characteristics and electron temperature. The optical diagnostics are emission and absorption spectroscopy. Optical methods can measure electron and particle densities in the gas phase of solution plasma. The densities of the neutrals and radicals can be measured by absorption spectroscopy in both the gas and solution phases. The diagnostics help to adjust plasma properties to get the most suitable conditions to obtain the desired functional material.

2.1 Optical Emission Spectroscopy

Plasma emission, known as Optical Emission Spectroscopy (OES), is one of the most remarkable and oldest diagnostic methods in plasma physics. The radiation emitted by atoms, molecules, and ions binds to plasma processes and parameters, allowing real-time observation. The OES is an easy experiment, especially in the visible and UV optical range. The method is non-invasive, i.e., it does not affect the plasma. Although the spectra are easy to obtain, interpretation can be quite complex, especially at low temperatures and pressure where plasmas are far from thermal equilibrium. The OES of the solution plasma in organic solvents is also difficult to interpret due to the formation of multiple radicals. Plasma in liquids has been intensively investigated (Bruggeman et al. 2009 and Bruggeman et al. 2010). The studies of solution plasma started in 2008 (Takai 2008).

Plasma spectroscopy is divided into passive emission spectroscopy and active absorption spectroscopy. In emission spectroscopy, the fundamental processes are the excitation of particles (atoms, molecules, ions) by inelastic collision with electrons, which produces excitation from lower-level l to upper-level u and de-excitation from u to l levels. Sometimes, de-excitation at an intermediate level k occurs by spontaneous emission with the probability of transition A_{uk} resulting in the emission of line $\varepsilon_{\mathrm{uk}}$. In the case of absorption spectroscopy, excitation from lower-level l to upper-level u occurs in the resonant electromagnetic field (i.e., by absorption with the probability of absorption B_{lu}), resulting in a decrease in the intensity of the electromagnetic field.

The emission intensity is correlated with the particle density in the excited state n_{u}. In contrast, the absorption signal correlates with the particle density in the lower state n_{l}, usually the ground state. Thus, the densities of the particles in the ground state are directly accessible by absorption spectroscopy; however, absorption techniques require much more experimental effort than emission spectroscopy. Since some principles of emission spectroscopy also apply to absorption and emission spectroscopy provides a variety of plasma parameters, we will first focus on emission spectroscopy, which is a very convenient passive diagnostic tool.

The two axes of a spectrum are wavelength (x-axis) and intensity (y-axis) (Fig. 13) (Frantz 2006). The central wavelength of an emission line at λ_0 is given by the photon energy $E = E_u - E_l$ corresponding to the energy difference of the electron transition from the top level to the bottom level $\lambda_0 = \frac{hc}{(E_u - E_l)}$, where h is the Planck constant, and c is the speed of light. E_{u} and E_{l} are the energies of the upper and lower levels, respectively. Because the energy of a transition is a characteristic of the particle species, the central wavelength is an identifier for particle radiation unless the Doppler effect shifts the wavelength. In principle, the wavelength axis λ_0 is easy to calibrate and analyze. Line intensity is quantified by the line emission coefficientε_{ul}:

$$\varepsilon_{ul} = n_u A_{ul} \frac{hc}{4\pi \lambda_0} = \int_{line} \varepsilon_\lambda d\lambda$$

where ε_{ul} unit is $\mathrm{W(m^2 sr)^{-1}}$, 4π is the detection solid angle measured in sr. The line profile P_λ connects the line emission coefficient with the spectral line emission ε_λ as:

$$\varepsilon_\lambda = \varepsilon_{ul} P_\lambda \text{ and } \int_{line} P_\lambda d\lambda = 1$$

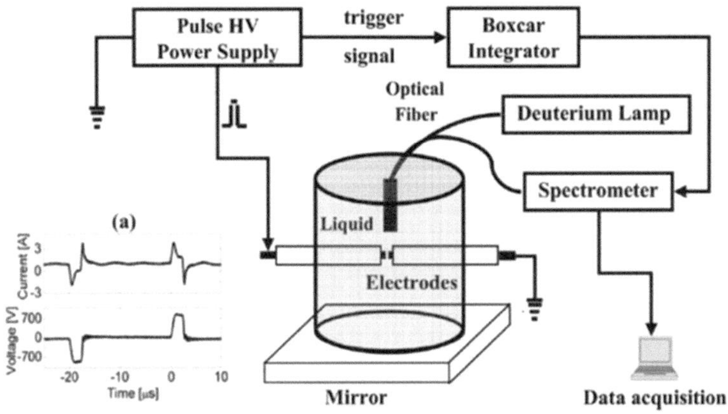

Fig. 13 Spectral line emission and its characteristics. G means the Gaussian shape of the line, and L represents the Lorentz shape of the line. ω_G and ω_L represent the full width at half maximum (FWHM) of the Gauss and Lorentz shapes spectral lines, respectively

The line emission coefficient is connected to the absolute line intensity I_{ul} in unit photons $(m^3\ s)^{-1}$:

$$I_{ul} = n_u A_{ul} \tag{7}$$

In OES, the line emission intensities depend on the number densities of excited species at the upper level, which rely on the plasma parameters, such as temperature and other species densities.

According to Eq. 7., the absolute intensity of a transition is directly correlated with the population density of the species on the excited level. The population density of the atoms or the molecules in excited states may obey the Boltzmann distribution, provided that the excited states are in thermal equilibrium between them. However, low-temperature solution plasma is non-equilibrium plasma, so population density in excited states does not obey the Boltzmann distribution. Consequently, the population density in the excited states depends not only on the temperature of the electrons but also on various plasma parameters: temperatures and densities of electrons and heavy particles and radiation field.

Thus, theoretical models are needed to describe the population density and consider the processes of population and depopulation for each level. Evidently, the number of processes is incalculable for molecules with vibrational and rotational levels and must be simplified somehow.

It should always be noted that plasma emission spectroscopy provides only a general understanding of the average parameters. The basic formula for the quantitative analysis of the line radiation is given by rearranging the equation:

$$I_{ul} = n_e n_n X_{ul}^{eff} (T_e, n_e, \ldots)$$

where n_n is the neutral density, and X_{ul}^{eff} is the excitation rate coefficient, which depends on plasma parameters. The emission rate coefficient is calculated from either the Collisional Radiative model or the corona model, which is the convolution of the direct cross-section with electron energy distribution function (Fantz 2006).

OES is the most common diagnostic method in the SP. Emission spectroscopy identifies plasma species (atoms, molecules, radicals, and ions) if the species emit radiation in the studied spectral range. The position of the spectral lines is a unique pattern that presents unambiguous evidence of the plasma species. The chemical reactions in SP can be established from the presence of various species. The OES experimental arrangement is straightforward (Fig. 14) (Bratescu et al. 2011a, b, c). The spectra identification is performed using databases, like the NIST database for atoms and atomic ions, and Pearse and Gaydon 1950, for molecules, molecular ions, and radicals.

OES is recorded using an optical fiber protected with a quartz window and connected to a spectrograph. The spectral measuring range is UV, visible, and sometimes near IR. In many experiments, the optical fiber was introduced in a quartz tube and ended with a quartz window to protect the optical fiber from gases and corrosive solutions. Long optical fibers are used to obtain a high signal-to-noise ratio, i.e., to set up the detection devices far from the electrically noisy SP system. The high voltage and current probes record the voltage and current waveforms on an oscilloscope.

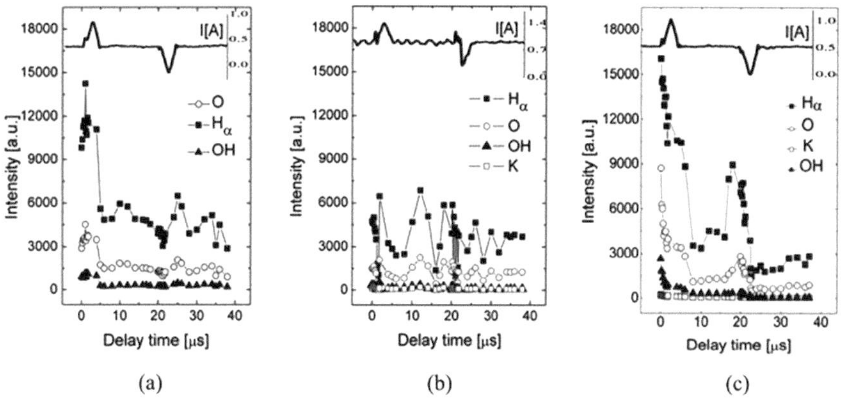

Fig. 14 The experimental arrangements for OES and electrical properties measurement of solution plasma. OES is recorded using an optical fiber protected with a quartz window and connected with a spectrograph. High voltage probe and current probe record the voltage and current waveforms on an oscilloscope, (reprinted from Bratescu et al. 2011a, b, c Copyright with permission from Elsevier Publishing)

Table 1 Hydrogen atom spectral data. NIST database

Hydrogen line	gA/s^{-1}	E/cm^{-1}	λ/nm
α	7.94×10^8	97,492.30	656.30
β	2.69×10^8	102,823.90	486.10

2.2 Electron Excitation Temperature Using the Balmer Hydrogen Lines

Molecule dissociation forms the hydrogen atom in aqueous and organic compound solutions. The electron collisions mainly produce the excitation of hydrogen atoms to different levels. A fast electron collides with a hydrogen atom, giving energy to the atom. The received energy by the atom corresponds to the excitation energy. The electron excitation temperature T_{exc} is evaluated by measuring the relative emission intensities of various transitions of the Balmer hydrogen lines (Balmer spectrum lies in the UV–visible range). The intensity I_α corresponds to the transition from the level with the principal quantum number 3 to 2. Assuming a Boltzmann distribution of the population of the atomic levels, I_α is $I_\alpha \approx exp\left(-\frac{E_\alpha}{T_{exc}}\right)$. The transition from the level with the principal quantum number 4 to 2 has an intensity I_β, which is $I_\beta \approx \exp\left(-\frac{E_\beta}{T_{exc}}\right)$. E_α and E_β are the energies of levels 3 and 4, respectively. The calculation of the electron excitation temperature is done using the below equation:

$$kT_{exc} = -\frac{E_\alpha - E_\beta}{\ln\ln\frac{I_\alpha A_\beta g_\beta \lambda_\alpha}{I_\beta A_\alpha g_\alpha \lambda_\beta}} \tag{8}$$

where E is the energy level of the hydrogen transition, g is the statistical weight, A is the Einstein transition probability, and λ is the wavelength. Data for hydrogen lines are listed in Table 1. (NIST database).

2.3 Electron Density from the Stark Broadening of the Balmer Hydrogen Lines

The shape and width of the spectral lines emitted by plasma are determined by collision processes perturbing the emitting atoms and molecules. The spectral line profile is a Voigt function characterized by broadening ω_V. The Voigt profile results from the convolution of a Gaussian function with a Lorentzian function, having the full width at half maximum (FWHM) ω_G and ω_L, respectively. The widths of the spectral lines are related to the values of the electron density and plasma temperatures. Thus, under the conditions of SP at atmospheric pressure, the FWHMs of the lines H_α, H_β, and H_γ result from the convolution of two Gaussian broadenings, Doppler (ω_D) and instrumental (ω_I), and two Lorentzian broadenings, the Stark (ω_S) and the van der Waals (ω_W). The natural broadening can be neglected in calculating the electron density for the solution plasma. The Heisenberg uncertainly principle explains the natural broadening because the atomic and molecular levels have a finite lifetime. The lineshape is a Lorentz function. The natural broadening is very small, of the order of 10^{-5} nm (Demtröder 2008).

The electron density measurement from the Stark broadening of the Balmer hydrogen lines works well if n_e values are arround 10^{20} m^{-3}. The n_e calculation consists of the deconvolution of the spectral profile extracting the van der Waals, Doppler, and instrumental broadenings. Using the computer simulation (Gigosos et al. 2003), FWHM expressions are given for the H_β Stark line profile and full width at half area (FWHA) for the H_α, H_β, and H_γ Stark profiles.

The Doppler broadening of the spectral lines is produced by the thermal motions of atoms and molecules and has a Gauss shape. The FFWHM of Doppler broadening is as

$$\omega_D = 7.16 \times 10^{-7} \lambda \left(\frac{T_g}{m} \right)^{1/2} \tag{9}$$

where T_g is the temperature of the emitters (gas temperature) in K, m is the atomic mass of the emitters in a.m.u., and λ is the central wavelength in nanometers. The instrumental broadening ω_I can be experimentally measured. Therefore $\omega_G = \omega_D + \omega_I$.

The SP in the gas phase operates at atmospheric pressure and room temperature. The van der Waals broadening is due to the collisions between the hydrogen-excited atoms with neutral atoms, molecules, and radicals. The degree of ionization in SP is smaller than 10^{-4}, so the neutral density is high. The van der Waals broadening cannot be neglected. The *FWHM* of van der Waals broadening expressed in nanometers is:

$$\omega_W \approx 8.18 \times 10^{-5} \lambda^2 \left(\alpha R^2 \right)^{2/5} \left(\frac{T_g}{\mu} \right)^{3/10} N \tag{10}$$

where N is neutral density in cm^{-3} and μ is the reduced mass of hydrogen and neutrals in a.m.u, the parameter α is hydrogen (emitter) polarizability in cm^{-3} and can be found in tables (Gigosos et al. 2003). R is the difference FWHM corresponding to the upper and lower levels.

In practice, from the experimentally measured FWHM of the Voigt profile ω_V, the Lorentz FWHM, ω_L can be calculated using the formula:

$$\omega_V = \left[\left(\frac{\omega_L}{2} \right)^2 + \omega_G^2 \right]^{1/2} + \omega_L.$$

The electron density can be extracted using different hydrogen emission lines (Gigosos et al. 2003). The computer fitting program suggests using the FWHA of the Stark profile line for a more precise estimate.

$$FWHA_L^\alpha = 0.549 \times \left(\frac{n_e}{10^{23}}\right)^{0.67965} \text{ for } H_\alpha$$
$$FWHM_L^\beta = 4.800 \times \left(\frac{n_e}{10^{23}}\right)^{0.68116} \quad (11)$$

and

$$FWHA_L^\beta = 1.666 \times \left(\frac{n_e}{10^{23}}\right)^{0.68777} \text{ for } H_\beta \quad (12)$$

$$FWHA_L^\gamma = 2.433 \times \left(\frac{n_e}{10^{23}}\right)^{0.68575} \text{ for } H_\gamma \quad (13)$$

where n_e is in m^{-3}, FWHA and FWHM are in nm.

2.4 Absorption Spectroscopy of Solution Plasma Gas Phase

The absolute number density of the molecules and the molecular radicals in the gas phase is measured by broadband absorption spectroscopy in the plasma gas phase (BAS-gas) (Luque et al. 2003 and Goyette et al. 1998).

In the case of BAS-gas measurements, the species number density of the molecular radicals and the molecules from plasma gas was calculated using the integrated absorption (A) over the band, knowing the transition oscillator strength from databases:

$$A = \int_{\text{absorption band}} \frac{I_0 - I(\nu)}{I_0} d\nu = \pi r_e \cdot f_{v'v''} \cdot n \cdot l \quad (14)$$

where I_0 and $I(\lambda)$ are the light intensity before and after passing the sample, respectively, $f_{v'v''}$ is the oscillator strength of the transition, and n is the species number density in cm^{-3} ($= cN_A$ with N_A being the Avogadro number). The symbols v' and v'' represent the vibrational quantum numbers. The absorption length l is expressed in cm.

A practical formula using wavelength is

$$A = 8.85 \times 10^{-20} \cdot \lambda_0^2 \cdot f_{v'v''} \cdot n \cdot l \quad (15)$$

where λ_0 is the average band wavelength in nm.

The typical experimental arrangement of the absorption spectroscopy is illustrated in Fig. 15. A UV–visible bifurcated optical fiber, optical components (lens and mirrors), a UV–visible broadband light source, and a spectrograph have to be employed (Bratescu et al. 2019). The light from the broadband source coupled with the bifurcated optical fiber passes through the solution and plasma gas, is reflected

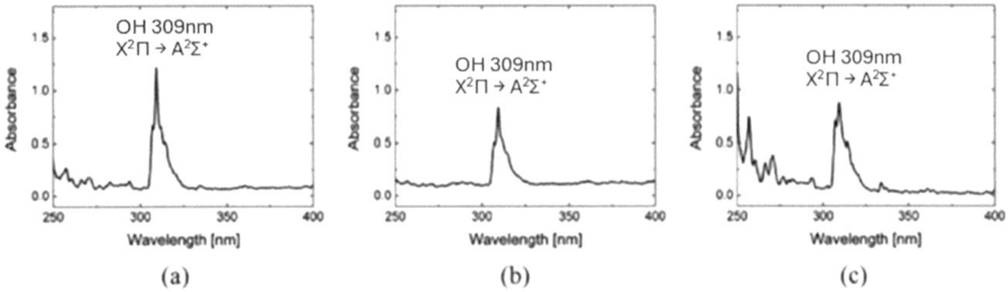

Fig. 15 A typical solution plasma experimental setup and the schematic arrangement of BAS-gas. Light from the broadband source passes through the solution and plasma gas, reflects on the mirror, and goes into the fiber to the spectrograph. **a** Construction of the optical absorption probe for the broadband absorption spectroscopy in solution (BAS-sol) method. The quartz tube surrounding the fiber ended with a window coupled with a stainless-steel mirror spaced apart at 5 mm from the window. In this space, the solution penetrates and is analyzed. The SP reactor was made from quartz to allow UV measurements (reprinted from Bratescu et al. 2019 Copyright, with permission from Wiley Publishing)

on the bottom mirror, and returns to the optical fiber coupled with the spectrograph. The bifurcated optical fiber was introduced into a homemade quartz tube to protect the fiber.

The absorbance spectrum A was calculated considering the subtraction of plasma emission (Luque et al. 2003 and Bratescu et al. 2019).

$$-\ln\frac{I_{s+SP} - I_{SP}}{I_s} = A. \tag{16}$$

where I_{s+SP} is the light captured from the SP emission and the broadband source, I_{SP} is the only emission from solution plasma when the shutter of the broadband source is *off*, and I_s is just the emission from the broadband source. In the case of solution plasma diagnostics, the integration time must be milliseconds, corresponding to an average spectrum over thousands of positive pulses. The measurements must be repeated at least tens of times with new solutions.

2.5 Gas Temperature Calculated from Blackbody Radiation

In the UV-vis OES, broadband background radiation was detected due to carbon clusters, eroded material from the grounded electrode, and nanoparticles. The broad background is the blackbody radiation. Wien's displacement law gives the relation between the wavelength of the maximum radiation and the blackbody temperature:

$$\lambda_{\max} = \frac{b}{T_b} \tag{17}$$

where λ_{max} is the maximum of the broadband emission band, T_b is the blackbody radiation temperature, and b is the Wien's displacement constant, equal to 2.8977×10^{-3} m K.

2.6 Electron Temperature Using the Electric Probe

Plasma density means the concentration of ionized gas particles in a given volume. Plasma density describes the plasma state, ionization state, and electron density (n_e) in fully or partially ionized gases. The gas phase plasma of the solution plasma is the discharge at the atmospheric pressure in the gas formed from the surrounding liquid.

The particles from the gas phase, ions, electrons, neutrals, and radicals, differentiate by mass and temperature. The average velocity of species v in plasma is $v = \left(\frac{8k_BT}{\pi m}\right)^{1/2}$. Here, k_B is the Boltzmann constant, and T and m are the temperature and mass of the species, respectively. The electron density and ion density are equal (on average); this number, which is much less than the neutral density, is, in fact, plasma density. The average electron velocity is enormous compared with ions and neutrals due to the high temperature and low mass of electrons.

Consider a small electrically isolated substrate (probe) suspended in the plasma. Initially, the probe will be hit by electrons and ions with charge fluxes, i.e., current densities will be.

$$j_e = en_e v_e/4 \text{ and } j_i = en_i v_i/4$$

where j_e and j_i are the current density for electrons and ions, respectively, v_e and v_i are the average velocity of electrons and ions in the plasma, and n_e and n_i are the electron and ion density.

Because of $j_e \gg j_i$, the probe starts accumulating a negative charge and gets a negative potential relative to plasma. Consequently, the random motion of ions and electrons in the region will be disturbed. On the probe, electrons are repelled, and ions are attracted. Since the probe charges are negative, the electrons are rejected, and ions are attracted. Accordingly, the electron flux decreases, but the object negatively charges until the electron flux decreases by repulsion, just enough to equilibrate the ion flux. Plasma is electric field-free, except for around perturbation, as described above, and plasma is equipotential. This potential is named plasma potential V_p, or space potential. The floating potential V_f will be attained when the net detected current is zero on the probe and is associated with the isolated probe.

The following considers the probe inserted in the plasma and maintained at a potential V set by an external power supply (Fig. 16). A conducting wall at the ground potential (0 V) acts as a reference voltage and a return current path. The plasma potential V_p is defined to the ground. The flux and the electric current will be zero when the probe is set at the floating potential V_f.

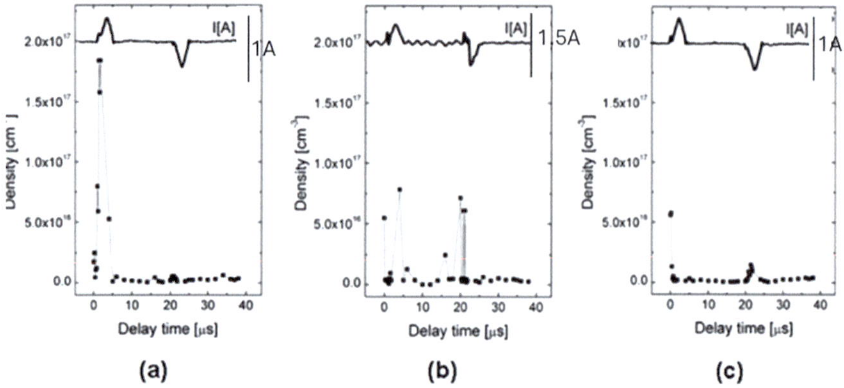

Fig. 16 The probe is inserted in the plasma and maintained at a potential V set by an external power supply. A conducting wall at ground potential (0 V) acts as a reference voltage and a return current path. The plasma potential V_p is defined to the ground (SP refers to solution plasma)

Figure 17 shows the dependence of the probe current density versus the probe voltage. By biasing the probe negatively to V_p, some electrons are prevented from reaching the probe. Still, the ion current density j_i remains at a value corresponding to the arrival rate of ions at the edge of the sheath and is limited to the random flux in the plasma.

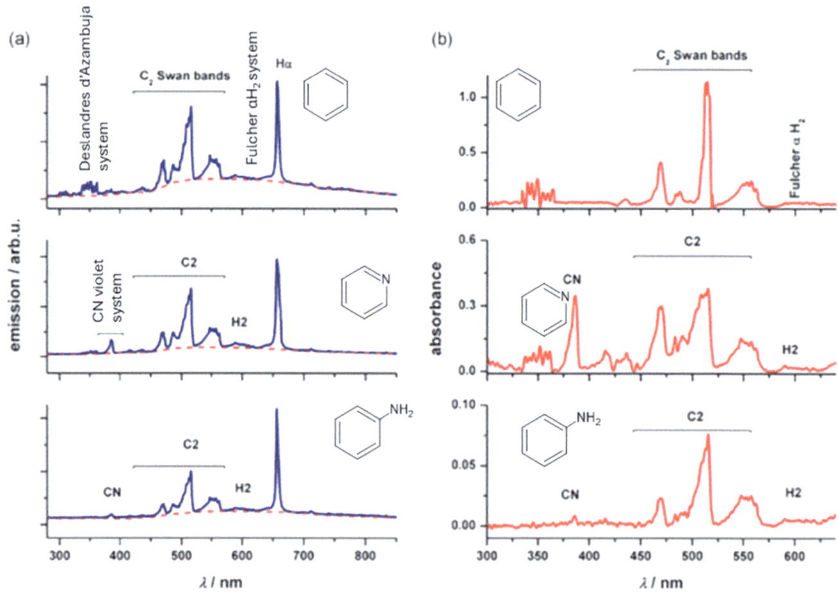

Fig. 17 The dependence of the probe current density versus the probe voltage

If, at V_f, the electrons drag is neglected, and the pressure gradient and the electric force per unit volume of electrons balance with each other (Clements and Smy 1973, 1974):

$$en_eE \approx \frac{d}{dy}(k_Bn_eT_e) \tag{18}$$

where T_e is electron temperature, the y-axis is perpendicular to the probe surface, and E is the electric field along the y-axis. Integrating the above equation from y = 0 (at the probe surface) to y = ∞ (the undisturbed bulk plasma), the floating potential V_f will be

$$V_f = \frac{k_B}{e}\left(T_e - T_{eprobe}\right) + \frac{k_B}{e}\int_{n_{eprobe}}^{n_e} T_e\frac{dn_e}{n_e} \tag{19}$$

The subscript "probe" is used for the probe surface, otherwise referred to as the gas-phase plasma.

Due to their equal mass, ions, and neutrals are assumed to be in equilibrium everywhere, and therefore, their temperatures will be just low in the vicinity of a cold probe. However, the temperature of the electrons may or may not be lower close to the probe.

When the electron temperature is the same at the probe and in the plasma bulk, the criteria for non-equilibrium between electrons and neutrals is

$$\frac{r_p}{\lambda_e} < \left(\frac{m}{m_e}\right)^{1/2} \tag{20}$$

where m is the ion mass, λ_e is the electron-neutral mean free path, and r_p is the probe dimension. If r_p is small enough

$$V_f = \frac{k_BT_e}{e}\ln\left(\frac{n_e}{n_{eprobe}}\right) \tag{21}$$

The gas phase in solution plasma is at atmospheric pressure; therefore, the sheath is very thin. With a cold probe inserted in plasma, the neutrals close to the probe (i.e., within a distance r_p) will be cooled by thermal conduction. The current of the ions collected by the probe is:

$$I_i = \frac{16\pi}{3}r_p\mu_in_ek_BT_e\left(1 - \frac{A}{1 - \frac{T_{probe}}{T}}\right) \tag{22}$$

where μ_i is ion mobility in plasma, and $A = 1 - TprobeT3/2$. The dependence of the floating potential on electron temperature is:

$$V_f = \frac{k_B T_e}{e} \ln\left(\frac{3 r_p e \left(\frac{k_B T_e}{2\pi m_e} \right)^{1/2} \left(1 - \frac{T_{probe}}{T} \right)}{4\pi \mu_i k_B T_e A} \right) \quad (23)$$

The electron temperature in the gas phase can be calculated from the floating potential value.

2.7 Energy Per Pulse

Figure 15 shows the experimental arrangement for electrical properties measurement of solution plasma. The voltage and current waveforms are usually recorded using a high-voltage probe (P6015A; Tektronix, Beaverton, OR) and an electric current probe based on the Hall effect (PR30, LEM), respectively, on an oscilloscope. The injected energy per pulse (W_{pulse}) is calculated by integrating the product between the voltage and the current waveforms over the pulse duration.

$$W_{pulse} = \int_0^T I(t)V(t)dt \quad (24)$$

where I(t) and V(t) represent the temporal waveforms of the current and voltage, respectively, and T is the time integration length.

2.8 In-Situ Absorption Spectroscopy of Solution

In the case of BAS-sol, a particular optical probe was constructed (Fig. 15a). The fused silica tube surrounding the fiber ended with a window coupled with a stainless steel (SUS) mirror spaced apart 5 mm from the window. The solution between the window and the SUS mirror is analyzed. The spectrograph was set in absorption mode measurement for the BAS-sol method. An optical fiber with a 5 m length couples with the spectrograph. The long optical fiber reduces the electrical noise from solution plasma and power supply during acquisition as much as possible (Bratescu et al. 2019).

From BAS-sol, the molar concentration (c) of the newly formed chemical compounds from the solution was calculated using the Beer-Lambert law of absorption (A), $c = \frac{A}{\varepsilon l}$, where ε is the molar extinction coefficient known from the database (Hirayama 1967). The reference was the spectrum of the broadband light source passing through the solution without plasma.

If integration is done over the broadband absorption, a deviation from the linearity will increase with increasing the analyte concentration. Generally, it is better to

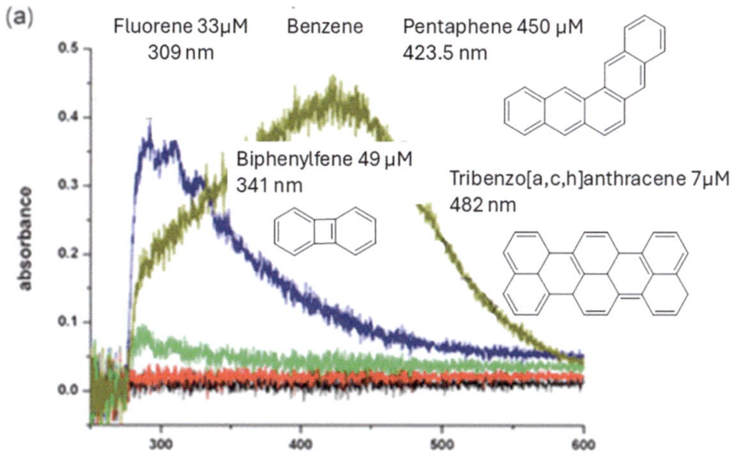

Fig. 18 Experimental setup CARS system used for solution plasma analysis of the solution. Two coherent lasers, OPO1 and OPO2, synchronized with the plasma pulse, are used to generate the anti-stokes Raman signal in the solution. The whole synchronization system is achieved using the signal generator Wave Factory 1956 from the pulse plasma power supply (reprinted from Bratescu et al. 2011b Copyright, with permission from AIP Publishing)

use a narrow wavelength range. However, if the measurements are over a small concentration range, the integration might be done over a band to increase the signal-to-noise level.

In-situ analysis of the solution during solution plasma operation can be performed by the Coherent Anti-Stokes Raman Spectroscopy (CARS) method (Bratescu et al. 2011b). The setup of CARS for the solution plasma is shown in Fig. 18. Two coherent lasers, OPO1 and OPO2, synchronized with the pulse voltage, are used to generate the anti-Stokes Raman signal in the solution. The whole synchronization system is achieved by using the signal generator from the pulse plasma power supply.

The principle of the method is illustrated in Fig. 19. Two photons from the pump laser OPO2 laser and one photon from the OPO1 laser, which corresponds to the Stokes Raman transition, generate the scattered anti-Stokes signal. The anti-Stokes signal is the optical beat of among these photon combinations. The optical beat of OPO1 and OPO2 lasers frequencies corresponds to a vibrational transition of the investigated molecule. The change of the molecular vibrations can be monitored by the *in-situ* CARS method in the solution.

Solution analysis has been performed using *ex-situ* absorption spectroscopy. Samples from the solutions from solution plasma were collected in time and analyzed by a spectrophotometer. The same solutions collected in time were also analyzed by liquid chromatography.

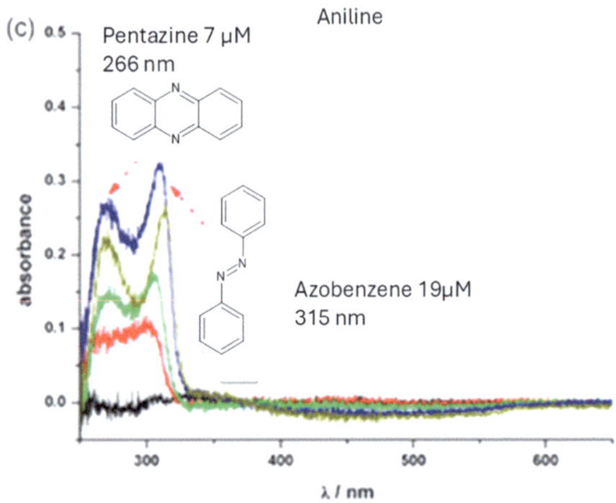

Fig. 19 The principle of CARS. The anti-stokes signal is the optical beat among the photons from OPO1 and OPO2 frequencies combinations. The optical beat of OPO1 and OPO2 lasers frequencies corresponds to a vibrational transition of the investigated molecule. The momentum and energy conservation laws of the photons must be fulfilled

2.9 Examples of Experimental Diagnostics of Solution Plasma

2.9.1 Optical Emission from Solution Plasma

Space and time-integrated OES experiments were performed. Different species (atoms, molecules, and radicals) exist in the gas phase in the solution plasma produced in pure water. The experiment was done with two types of electrode materials: W and Ta. Figures 20 and 21 show the OES together with the identification of the species (Miron et al. 2010).

The hydroxyl radical (OH) emission attracted much attention due to the differences in the spectral profile in the circulated water and the not-circulated SP in water when W and Ta electrodes were used. The band (0, 0) of the OH radical (Q1 branch forms the peak at 307.8 nm) was detected to be the strongest line among other OH radical bands when the W electrodes were used, while for the Ta electrodes, the head of the band (1, 1) (R2 branch peaks at 312.6 nm) was the strongest. Different temperatures of the solution plasma could explain the shift in the maximum intensity of the OH bands in solution plasma with W and Ta electrodes.

The evolution of OH radical was studied in an exciting experiment in the solution plasma, where W and SUS electrodes were used in a bridged (small inter-electrode) distance and unbridged (long inter-electrode) distance between W and SUS electrodes (Baroch et al. 2011).

The OES analysis shows the hydrogen peroxide production and the erosion of needle electrodes for the solution plasma in water with a pulse energy of about 10 mJ depending on the plasma type (connected or bridged discharge mode, i.e., small inter-electrode distance and separated or unbridged discharge mode, i.e., large inter-electrode space), different electrode materials (tungsten and stainless steel), and the solution electrical conductivity (500, 750, 1000 and 1250 μS/cm). The intensity of

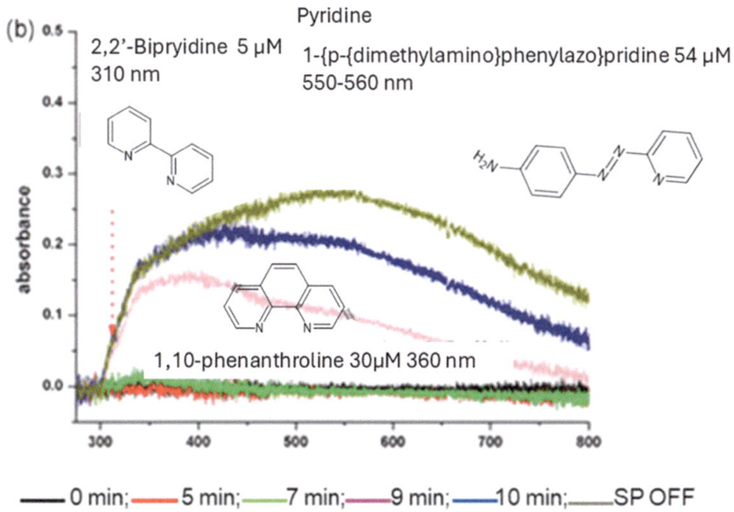

Fig. 20 The OES in SP circulated water using W (**a**) and Ta (**b**) electrodes. When a positive voltage pulse was applied, the emission spectra were taken in the same experimental conditions (discharge voltage V = 1,090 V, pulse frequency 25 kHz, pulse width 4 μs, water flow rate of 250 ml/min) the plasma. In figure (**a**), the small arrows mark the position of the H_2O and the W emission lines at 662.8 and 333.9 nm, respectively. The weak emission band of the O_2 Chamberlains airglow system was magnified and displayed for the W electrodes (**c**), with the band head at 400.9 nm, and for the Ta electrodes (**d**), with the band head at 403.1 nm (reprinted from Miron et al. 2010 Copyright, with permission from Springer Publishing)

the radiation emitted by the OH hydroxyl radical is high in the case of unbridged mode, with dependence on the electrode material or the conductivity of the solution. The hydrogen peroxide production did not depend on the electrode material nor correlated with the OH radical emission line intensity (Fig. 22).

The OES analysis shows the hydrogen peroxide production and the erosion of needle electrodes in the SP in water with a pulse energy of about 10 mJ depending on the plasma mode (connected or bridged discharge mode, i.e., small inter-electrode distance and separated or unbridged discharge mode, i.e., large inter-electrode space), the electrode material (tungsten and stainless steel), and the solution electrical conductivity (500, 750, 1000, and 1250 μS/cm). The intensity of the radiation emitted by the OH hydroxyl radical is high in the case of unbridged mode, depending on the electrode material and the solution conductivity. The hydrogen peroxide production in the solution did not depend on the electrode material and was correlated with the OH radical emission line intensity (Fig. 22).

In an experiment to elucidate the fastest formation routes of nanocarbons in the SP, the OES showed the relation between the excited states of molecules, radicals, and atoms in the gas phase and the obtained solid products. Figure 23 represents the OES spectra of the SP gas phase formed in hexane, hexadecane, cyclohexane, and benzene. For all OES spectra, H_α, H_β, and C_2 lines and bands were detected. The ratios of the peak intensities were almost the same.

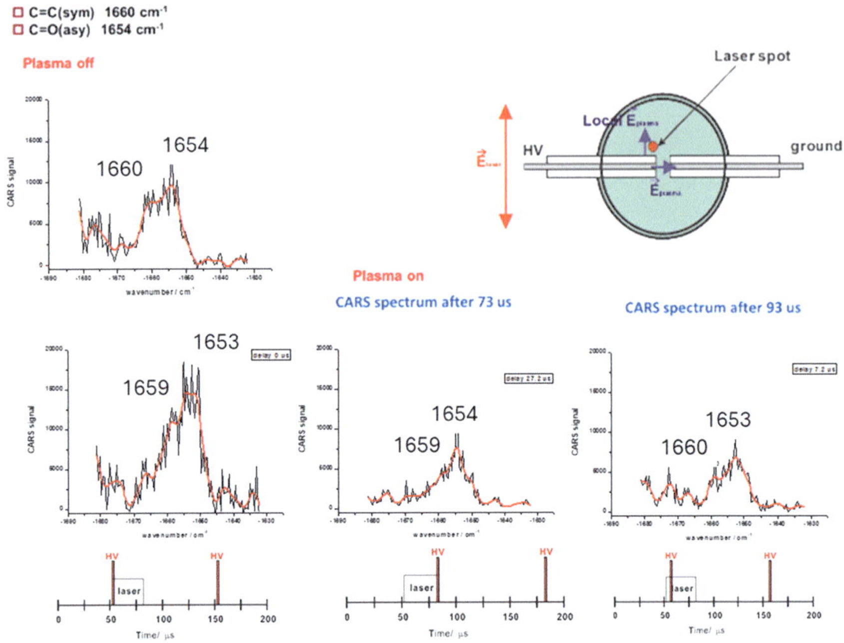

Fig. 21 The OES in not-circulated water using W (**a**) and Ta (**b**) electrodes. The emission spectra were taken in the same experimental conditions (discharge voltage V = 540 V, pulse frequency 25 kHz, pulse width 4 µs) when a positive voltage pulse was applied to the plasma. In figure (**a**), the small arrows mark the position of the H_2O and the W emission lines at 662.8 and 333.9 nm, respectively. The weak emission band of the O_2 Chamberlains airglow system was magnified and displayed for the W electrodes (**c**), with the band head at 400.9 nm, and for the Ta electrodes (**d**), with the band head at 403.1 nm (reprinted from Miron et al. 2010 Copyright, with permission from Springer Publishing)

Moreover, the background shapes were roughly the same shape and intensity. The background was associated with the heated particles inside the SP gas phase. According to Wien's displacement law, from the background of the blackbody radiation, the temperature was calculated to be 4000–5000 K. Such temperatures are sufficient to decompose the organic molecules and form graphite.

The SP is widely used in nanomaterial fabrication. The experiment synthesizing the iron oxide and oxide-hydroxide nanoparticles using the SP method showed that 80 nm size, rice-shaped morphology can be easily obtained with low plasma energy in the glow discharge mode (Fig. 24). An oxidation rate of 0.13 µg mL^{-1} s^{-1} and a reduction rate of 0.04 µg mL^{-1} s^{-1} of iron ion in the SP were measured when the input plasma energy per pulse was 0.4 mJ, during iron oxide and oxide-hydroxide nanoparticles synthesis. In the liquid phase, the redox reaction rates depend on the dissociation and excitation processes in the plasma gas phase (Fig. 24c). It was observed that the changes in the relative number densities of oxygen and hydrogen

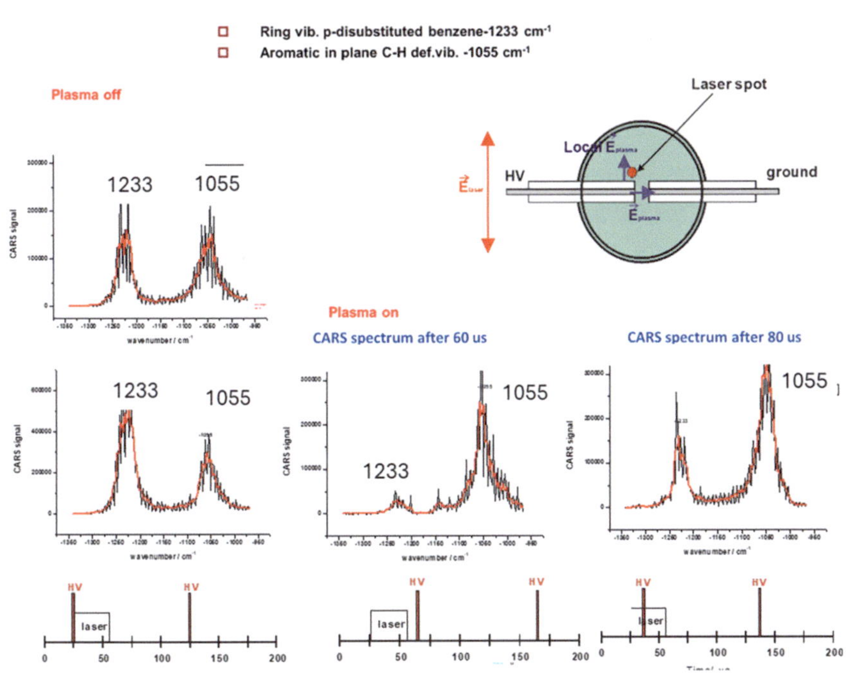

Fig. 22 Spatially and temporally integrated emission spectra for the discharge using tungsten (**a**) and stainless steel (**b**) needle electrodes in the case of connected (bridged) and separated (unbridged) discharge modes (frequency 10 kHz, pulse duration 4 μs, and σ = 1000 μS/cm) (Baroch et al. 2011)

Fig. 23 The OES for plasmas formed in hexane, hexadecane, cyclohexane, and benzene (Morishita et al. 2016)

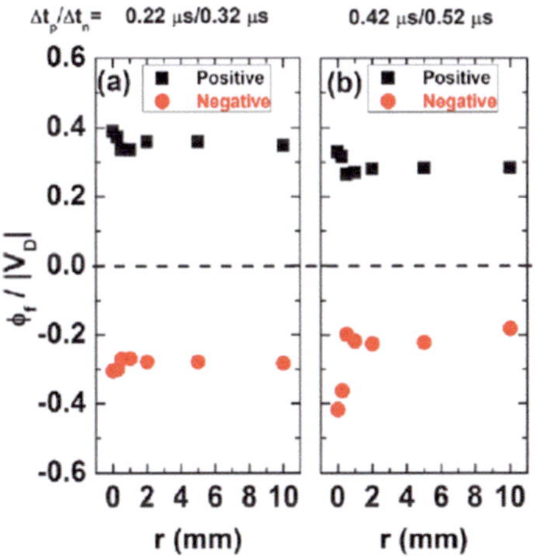

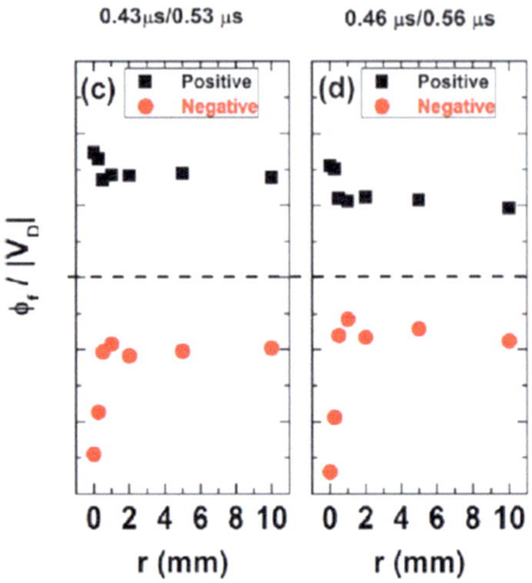

Fig. 24 The OES of solution plasma **a** at 0.18 mJ/pulse characteristic to a glow discharge and **b** at 0.44 mJ/pulse characteristic to arc discharge (both spectra **a** and **b** have been represented at the same ordinate); **c** dependence of the relative atomic oxygen and hydrogen number densities on energy per pulse (left); Dependence of the ratio between the emission line intensities of oxygen and hydrogen on the energy per pulse (right) (reprinted from Bratescu et al. 2011a, b, c Copyright, with permission from Elsevier Publishing)

atoms depending on the plasma-operating regime are connected with the oxidation and the reduction rates in the liquid phase.

The SP characteristics and operation regime were correlated with the structural properties of the iron oxide and oxide-hydroxide nanoparticles, which were more effortless and rapidly synthesized by the SP method than by other methods.

2.9.2 Electron Density in Solution Plasma

In the SP, n_e is known from the Stark broadening of hydrogen emission lines. Size-controlled gold nanoparticles have been synthesized using SP (Bratescu et al. 2011a). The gold nanoparticles exhibit sizes of 1, 2, and 10 nm when the solution pH was adjusted from 12 to 3. The chemical environment surrounding the gold nanoparticles depends on the preparation conditions and determines the electrostatic interaction among the nanoparticles. Plasma diagnostics shows that a high electron density generates a more significant number of hydrogen radicals, which represent the main component in the reduction process of the gold ion into the neutral form.

In the gold nanoparticle synthesis process, plasma supplies H radicals, which are used to reduce the gold ion to the neutral atom without any other chemical reducing agent. The relative number of the H radical depends on the electron number density because these are mainly produced by electron impact dissociation of water molecules. The relative amount of gold atoms present in the plasma gas phase is not influenced by the solution pH. In a high pH solution, part of the H radicals recombines with OH radicals.

Various structural properties of the synthesized materials correlate with the SP characteristics. Figure 25a. shows the electron density (n_e) dependence on the solution conductivity. However, the solution conductivity depends on the solution pH. In such a situation, the size of the gold nanoparticles was controlled by simply adjusting the pH solution.

In the paper mentioned above, the Stark broadening of the H_β line was used to calculate n_e. The gas temperature was 1600 K. The n_e calculation needed the spectral deconvolution procedure for the H_β line, having the van der Waals, Doppler, and instrumental broadenings to be 0.03, 0.02, and 0.02 nm, respectively (Laux et al. 2003, Bruggeman et al. 2009, and Yubero et al. 2005). In Fig. 25, the dependence of n_e on the solution conductivity is shown together with an example of the H_β line (black curve) fitted with a Voigt line profile (red dashed curve), from which FWHM was extracted, and the electron density was calculated from the Eq. 12:

$$FWHM = 4.800 \times \left(\frac{n_e}{10^{23}}\right)^{0.68116}$$

Different sizes of gold nanoparticles were explained by the reduction reaction of gold ions by hydrogen. Different hydrogen concentrations determine different sizes of the nanoparticles (Fig. 25b).

Measuring the electron density in SP was essential in improving the power supply. In the low-pass filter circuit used for SP generation, the electron density and temperature are higher than in the SP produced with a conventional circuit, making the low-pass filter circuit better suited for stable material processing with SP (Heo et al. 2017).

2.9.3 Time-Resolved Radical and Electron Densities in Solution Plasma

The SP operates in pulsed mode. Therefore, the time-resolved OES in SP is necessary to describe the time evolution of radical and electron densities and their temperatures.

The OES of the reactive species formed in the gas phase were collected with an optical fiber immersed in water at 15 mm above the gap between the electrodes using a tube that ended with a quartz window. The optical fiber was connected to a 21 ns time resolution spectrograph (the optical resolution was 2.4 nm in the spectral range from 180 to 1100 nm). A 200 ns gate of a boxcar integrator device was synchronized with the pulsed power supply to collect the spectra at different time delays related to the pulse voltage. The time delay was changed from 0 to 35 μs, corresponding to a time over three pulses at a plasma frequency of 25 kHz. The time-resolved OES was acquired at different times over the positive and negative voltage pulses and the regions between the applied pulses. The experimental conditions were kept constant, except for the electrode materials (W and Ta) (Miron et al. 2010).

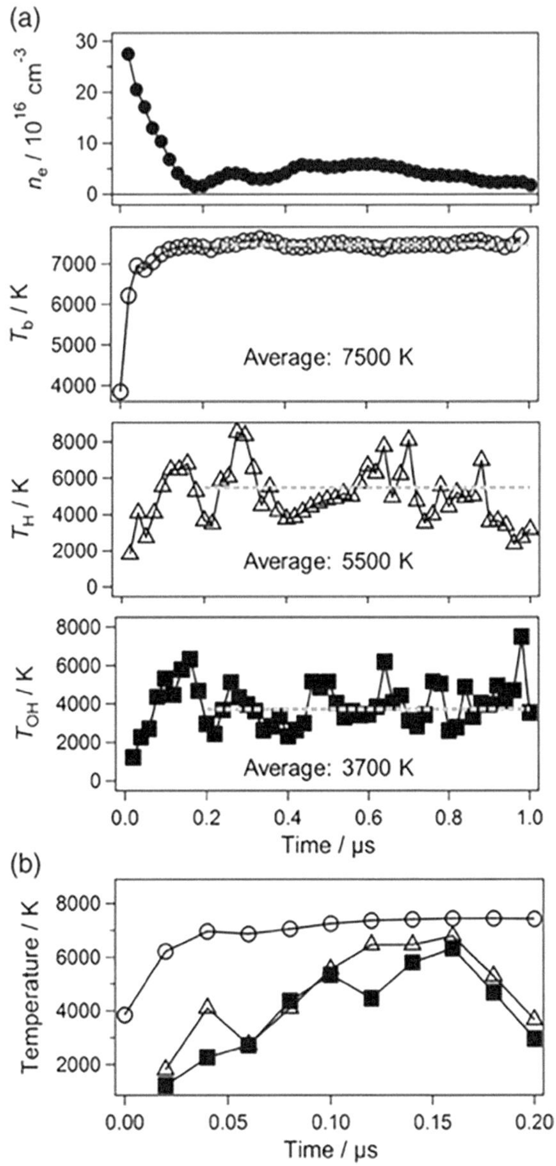

Fig. 25 **a** The dependence of n_e on solution conductivity. The insert in the figure represents an example of the H_β line (black curve) fitted with a Voigt line profile (red dashed curve), from where the full width at half-maximum (FWHM) was extracted, and the electron density was calculated. **b** Gold NPs size distributions were measured by small-angle X-ray scattering diffractometry corresponding to different initial pH (reprinted from Bratescu et al. 2011a Copyright, with permission from ACS Publishing)

The time evolution of the excited species depended on the electrode material used to generate the discharge but was independent of the pulse polarity of the applied voltage. The reactive species were identified in the emission spectra in a high concentration between the applied voltage pulses. The light emitted between pulses can be explained by a DC voltage component of about 9 A for W and 7 A for Ta electrodes (circulated water) and 2.5 A for W and 3 A for Ta electrodes (not-circulated water) (Fig. 26).

In the above-described SP experiment, two types of electrodes, W and Ta, were used in pure water, circulated and non-circulated. The intensity ratio of the H_α and H_β time-dependence was measured to calculate the excitation temperature. The radical species became almost steady-state values in the regions between the plasma pulses in both cases. Under the assumption of local thermal equilibrium, the mean calculated electron excitation temperatures were 0.5 ± 0.1 and 0.8 ± 0.2 eV for the W and Ta electrodes, respectively. Multiple electron collisions explain the low electron excitation temperatures, a plasma at atmospheric pressure.

A streak camera with nanoseconds time resolution, coupled with a spectrograph and time-synchronized with the power supply, was used to measure the time dependence of various parameters of solution plasma in NaCl aqueous solution (Fig. 27) (Banno et al. 2015).

The nanosecond processes in the plasma have been clarified by nanoseconds resolution OES. In the early time region after the onset of the plasma emission, increases in the nanosecond temperature of representative transient species were observed (rising stage). In the time region corresponding to the rising stage, the number density of the electrons is high. The rapid temperature increase results from the frequent collisions between the accelerated charged particles and the unexcited particles. The velocities of the temperature increases are almost identical among H and OH radicals, suggesting simultaneous generations of these radicals. After the rising stage, electron excitation temperature T_{exc} and rotational temperature of OH radical, T_{OH} fluctuate around their average values on the nanosecond timescale (fluctuating stage). The changes in velocities of the T_{exc} and T_{OH} were found to strongly correlate with n_e, indicating that the radical temperatures are affected by collision frequencies with the charged particles. The radical temperatures do not decay when n_e is more than 3 to 4×10^{16} cm^{-3}. n_e of more than 3 to 4×10^{16} cm^{-3} is necessary to sustain the solution plasma. The average T_{OH} is smaller than T_{exc}, indicating that OH radical loses its energy fast owing to collisions with H_2O molecules.

The time dependence of the absolute number density of various atoms, molecules, and radicals from the gas phase was obtained by Broadband Absorption Spectroscopy in the plasma gas phase (BAS-gas) in combination with a time-resolved setup (Fig. 28) (Miron et al. 2011).

For the first time, time-resolved BAS-gas was performed in the solution plasma. The electrical and optical characteristics of the solution plasma-initiated in hydrochloric acid (HCl), potassium chloride (KCl), and potassium hydroxide (KOH) aqueous solutions were investigated. The solution conductivity was adjusted before the solution plasma operation. The temporal evolution of the reactive species formed in the solution plasma was measured via time-resolved OES.

The time-resolved BAS-gas technique was employed to find the time evolution of the density of the hydroxyl radical in the ground state ($X^2\Pi$). The time evolution of the reactive species was measured during the positive and negative pulses and in the regions between the pulses. The time-emission intensities depended on the nature of the solution and the polarity of the applied pulse at the electrodes. The excited

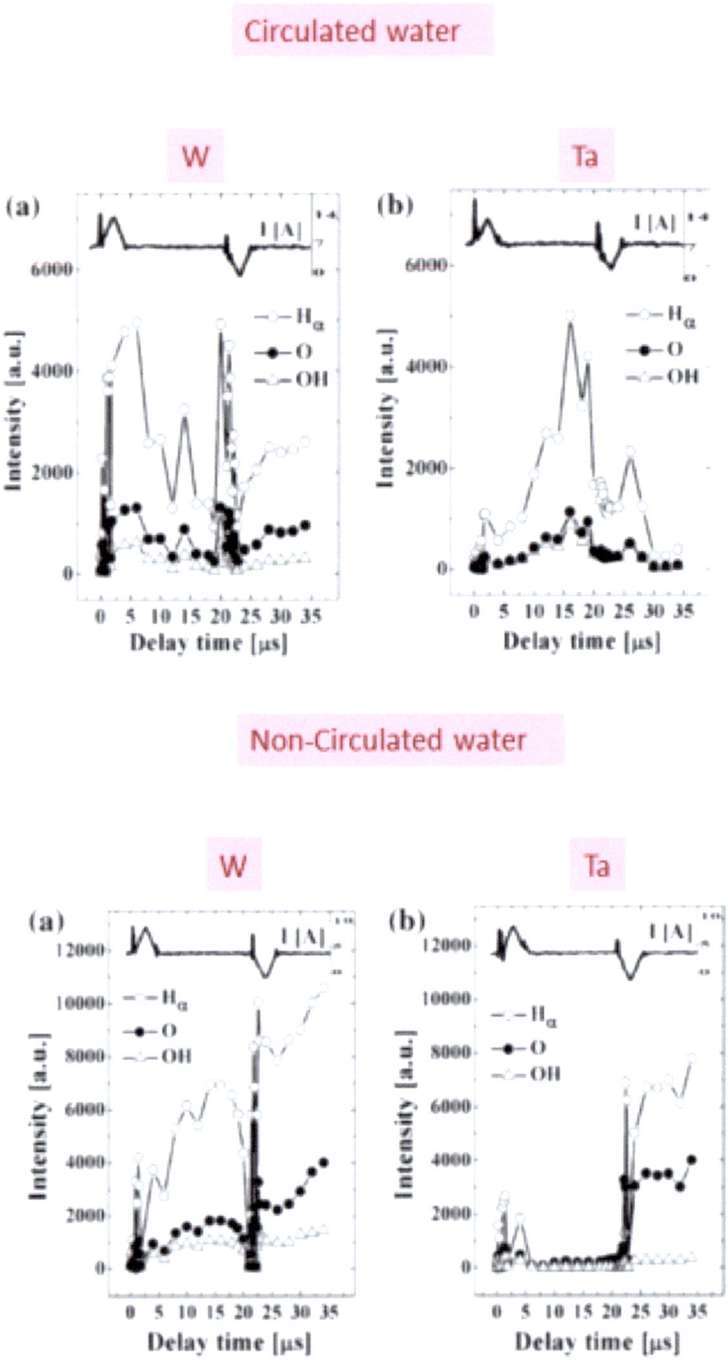
Circulated water
W
Ta
(a)
(b)
I [A]
Hα
O
OH
Intensity [a.u.]
Delay time [μs]
0 5 10 15 20 25 30 35
0
2000
4000
6000
Non-Circulated water
W
Ta
(a)
(b)
I [A]
Hα
O
OH
Intensity [a.u.]
Delay time [μs]
0 5 10 15 20 25 30 35
0
2000
4000
6000
8000
10000
12000

◂**Fig. 26** Top figure: time-resolved OES of the gas phase in SPP with circulated water using W (**a**) and Ta (**b**) electrodes in the same experimental conditions (discharge voltage 1090 V, frequency 25 kHz, pulse width 4 µs, water flow rate 250 ml/min). Bottom figure: time-resolved OES of the gas phase in SPP with non-circulated water using W (**a**) and Ta (**b**) electrodes in the same experimental conditions (discharge voltage 450 V, frequency 25 kHz, pulse width 4 µs) (reprinted from Miron et al. 2010 Copyright, with permission from Springer Publishing)

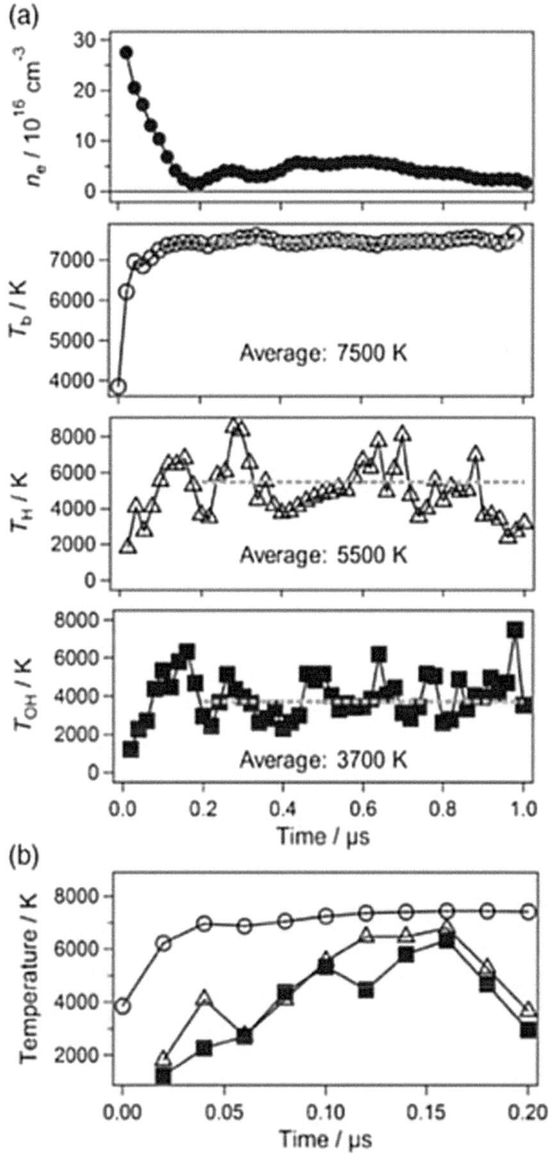

Fig. 27 Delay time dependence of n_e (electron density), T_b (blackbody radiation temperature), T_H (electron excitation temperature, T_{exc}), and T_{OH} (rotational excitation temperature. For T_b, T_H, and T_{OH}, the averages in the time range between 0.2–0.9 µs are indicated by gray dotted lines. **b** The initial stages for T_b, T_H, and T_{OH} are enlarged (reprinted from Banno et al. 2015 Copyright, with permission from IOP Publishing)

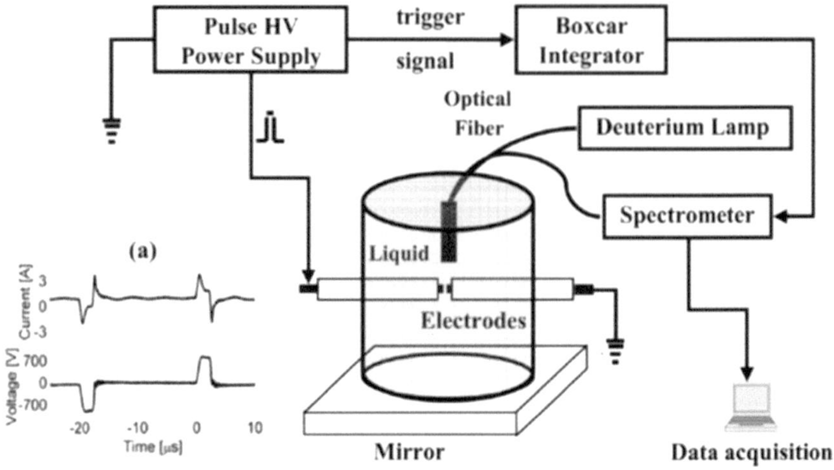

Fig. 28 The experimental setup of the time-resolved BAS-gas in the solution plasma in HCl, KCl, and KOH solutions. The voltage and current waveforms of the pulse power supply are included in the inset (**a**) (reprinted from Miron et al. 2011 Copyright, with permission from AIP Publishing)

species generation in the regions between voltage pulses (in the ranges of 4–20 μs and 24–35 μs) also can be explained by the DC voltage and current component of about 0.35 A for HCl, 0.83 A for KCl, and 0.75 A for the KOH solution which continuously sustains the solution plasma (Fig. 29).

Time-resolved BAS-gas of the OH radical $\left(X^2\Pi \rightarrow A^2\Sigma^+\right)$ corresponding to the vibrational quantum numbers 0–0 at $\lambda = 309$ nm) were acquired at various time

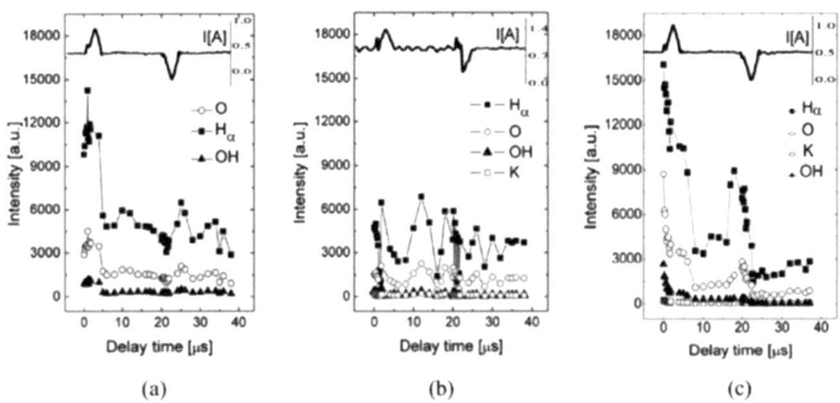

Fig. 29 Time-resolved OES of the solution plasma in **a** HCl, **b** KCl, and **c** KOH aqueous solutions using W electrodes, acquired under the same experimental conditions (discharge voltage 700 V, frequency 25 kHz, pulse width μs). Lines are used as guides for the eyes (reprinted from Miron et al. 2011 Copyright, with permission from AIP Publishing)

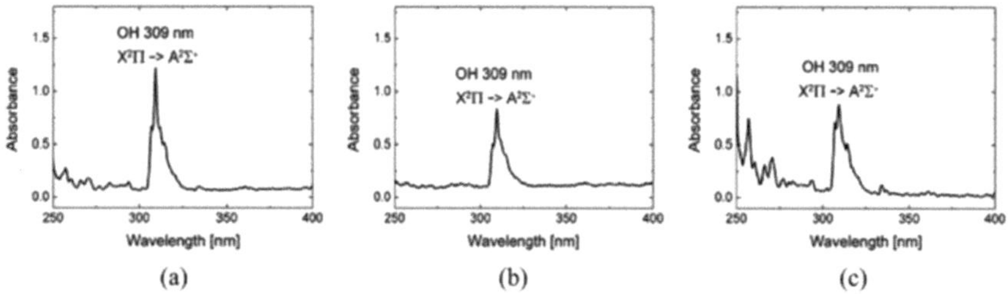

Fig. 30 BAS-gas of the OH radicals measured in **a** HCl, **b** KCl, and **c** KOH aqueous solutions, on the positive applied voltage, under the same experimental conditions (discharge voltage 700 V, frequency 25 kHz, pulse width 4 μs) (reprinted from Miron et al. 2011 Copyright, with permission from AIP Publishing)

delays relative to the positive voltage pulse in the 0–38 μs range. The experimental conditions for the solution plasma in HCl, KCl, and KOH solutions were a discharge voltage of 700 V, a frequency of 25 kHz, and a pulse width of 4 μs. (Fig. 30).

Figure 31 shows the time-resolved of the absolute number density of the OH radical in the ground state ($X^2\Pi$). The density depends on the applied voltage polarity and is low in the regions between the pulses for all three types of solutions. The absorption signal was not detected in these regions. The time-resolved OES (Fig. 30) shows that the emission intensity of the reactive species, including that of the OH radicals, is also low in these regions. The signal-to-noise ratio in the absorption spectra was found to be about 10. The OH radical density was highest for the HCl solution plasma (2×10^{17} cm^{-3}) when positive voltage pulses were applied to the electrodes. KOH is highly basic and can be an important source of hydroxyl radicals. However, the conductivity of the HCl and KCl solutions, with the highest OH emission intensity and high number density of OH radical in SP, was measured at a conductivity of about 10–80 μS/cm. At higher conductivities, the emission intensity and absorption, related to the number density on the ground state, became low due to the weakening of the electric field in a high-conductivity liquid. Therefore, the lower conductivity of the KCl and HCl solutions determined that the generation of the OH radicals in the ground state was higher than that in the KOH solution.

2.9.4 Radicals and Molecules Densities in the Gas and Liquid Phases of Solution Plasma

In a simple and original experiment, the radical and molecule number densities were measured in the solution plasma in aromatic organic molecules: benzene, pyridine, and aniline. The electron excitation temperature (T_{exc}) was calculated from the intensity ratio of the hydrogen emission lines at 486.1 and 656.3 nm, considering the emission intensities after the continuum background radiation subtraction. T_{exc} was correlated with the number densities of molecules and radicals found in

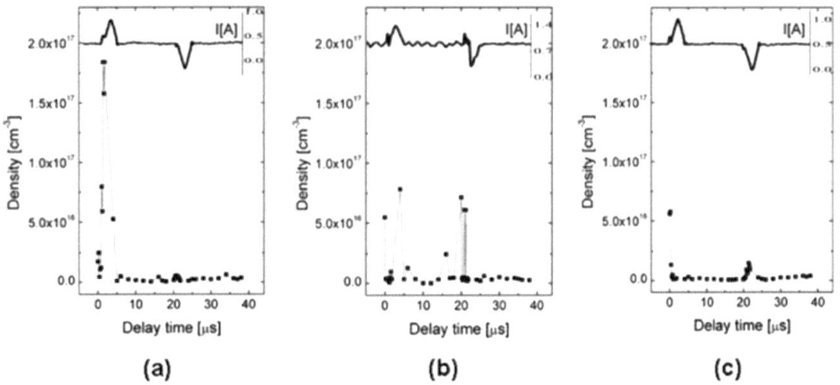

Fig. 31 Time-resolved OH radical, absolute number density measured in **a** HCl, **b** KCl, and **c** KOH solutions under the same experimental conditions (discharge voltage 700 V, frequency 25 kHz, pulse width 4 μs). Lines are used as guides for the eyes (reprinted from Miron et al. 2011 Copyright, with permission from AIP Publishing)

the gas phase. *In-situ* absorption spectroscopy of the solution from the SP was also investigated (Bratescu et al. 2019).

The solution plasma reactor was a quartz beaker with two lateral and opposite holes where silicon stoppers were used to fix the electrodes (Fig. 15). The electrodes were made from 1 mm diameter nickel wire covered with insulating ceramic tubes except for a small tip (~ 1 mm) from where the discharge occurs. The electrode gap distance was adjusted at 1 ± 0.1 mm. The solution plasma processed 30 ml of benzene, pyridine, or aniline for ten minutes using a bipolar pulsed power supply (usually MPP-HV04, Kurita Seisakusho). During plasma operation, the applied voltage between the electrodes was about 1.3 kV at a repetition frequency of 20 kHz and an average pulse width of 1.0 μs. The voltage and current waveforms were recorded using a high-voltage probe (P6015A, Tektronix) and an electric current probe (PR30, LEM), respectively, on an oscilloscope (Tektronix, TDS 2014C) (Fig. 32).

The high evaporated mass of benzene determines the most significant number density of C_2 in the plasma gas phase compared to pyridine and aniline. Furthermore, the slightly larger excitation temperature in the solution plasma in benzene compared to aniline contributes to small fragment formation by dissociation and, consequently, a greater density of radicals. The relative amount of C_2 to benzene, pyridine, and aniline gas molecules was estimated to be 8×10^{-4}, 3×10^{-4}, and 6×10^{-5}, respectively, considering the Avogadro number (Bratescu et al. 2019).

In the solution plasma in pyridine and aniline in Table 2, the CN radical number density was 1.1×10^{16} and 3.4×10^{14} cm^{-3}, respectively. The high value of the vaporization enthalpy and elevated boiling temperature require higher energy to vaporize and operate the solution plasma in aniline than pyridine. Consequently, a small number of aniline molecules evaporate and dissociate in the solution plasma. Therefore, the number density of CN and C_2 radicals in plasma gas is lower in aniline than in pyridine.

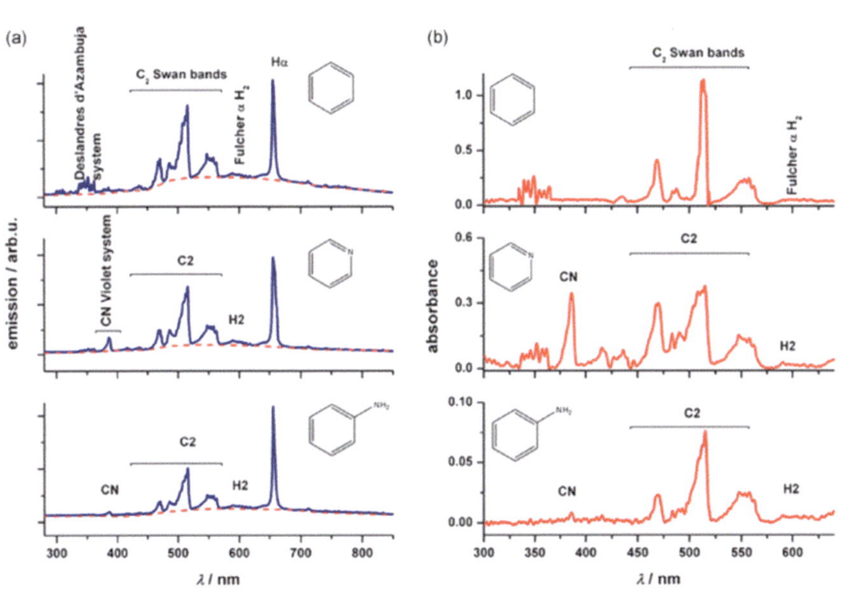

Fig. 32 **a** Emission and **b** absorption spectra obtained by BAS-gas. In the emission spectra, the dotted red line represents the blackbody radiation. The number densities of C_2 and CN radicals in the ground state and H_2 molecules in the electronically excited state were determined from the absorbance spectra (reprinted from Bratescu et al. 2019 Copyright, with permission from Wiley Publishing)

Table 2 Number density (cm^{-3}) of C_2 and CN radicals and H_2 molecules in the plasma gas phase measured by BAS-gas. The experimental conditions are given in the reference Bratescu et al. 2019

Radical	Oscillator strength	Benzene	Pyridine	Aniline
·C≡C·on the ground state $a^3\Pi_u$	0.027	$(2.2 \pm 1.3) \times 10^{16}$	$(6.8 \pm 4.1) \times 10^{15}$	$(1.5 \pm 0.9) \times 10^{15}$
·C≡N on the ground state $X^2\Sigma^+$	0.027	–	$(1.1 \pm 0.7) \times 10^{16}$	$(3.4 \pm 2.0) \times 10^{14}$
H_2 on the excited state $a^3\Sigma_g^+$	0.14	$(1.6 \pm 1.0) \times 10^{14}$	$(5.4 \pm 3.2) \times 10^{13}$	$(1.6 \pm 1.0) \times 10^{13}$

The blackbody radiation temperature was calculated from the broad emission background for all the experiments. Due to carbon clusters and nanoparticles, blackbody radiation was detected in all UV–vis OES measurements. The emission from the CH, NH radicals, and N_2 molecule in the plasma gas in aniline and pyridine was not observed, indicating that the dissociation initially occurs by breaking the carbon-hydrogen bond or carbon–nitrogen bond and then carbon bonds. Hydrogen molecules detected in spectra and measured as having approximately 2 to 16 × 10^{13} cm^{-3} number density on the electronically excited state at 12 eV from the ground state confirms the dissociation of hydrogen of molecules. In the case of benzene, the bond dissociation energy of the first hydrogen is 4.9 eV. Second, hydrogen dissociation occurs at a smaller energy range from 3.38 to 4.72 eV, depending on the position of the hydrogen. In this way, the phenyl radical is readily produced in the plasma gas benzene phase. In the case of pyridine, the minimum dissociation bond energy of the hydrogen atom is 4.55 eV. Aniline needs only 3.47 eV to break hydrogen from the amine group. For the solution plasma as a non-LTE, the electron energy can reach 10 eV in the tail of the electron energy distribution function.

The BAS-sol measurements show the molar concentration of the new compounds formed into the solution phase at different times in the solution plasma processed in the aromatic molecules. After more than 10 min, a solid carbon material is obtained, and the transmission of the solution decreases, making it difficult to measure the absorbance. Figure 33 shows *in-situ* BAS-sol results of the quantitative analysis in (a) benzene, (b) pyridine, and (c) aniline. The maximum molar concentration was calculated from the absorbance considering the molar extinction coefficient ε (Hirayama 1967).

The molar concentration of molecules formed in the solution phase ranges from 5 to 450 μM. CG-MS analysis confirmed the results obtained by BAS-sol.

In the CARS experiment in the solution plasma, the influence of the electric field of plasma on *p*-benzoquinone (*p*-BQ)_molecule vibrations was studied. By changing the delay time between the electric pulse of plasma and the laser pulse synchronous with the detection time, the ring vibration of the *p*-BQ molecule changes. The observed CARS signals of the vibrational transitions of the *p*-BQ solution in water were two bands at 1654 and 1660 cm^{-1}, corresponding to the C=O asymmetrical and the C=C symmetrical stretching vibrations, respectively (Fig. 34), and the bands at 1233 and 1055 cm^{-1} corresponding to the ring vibration of *p*-disubstituted benzene and the aromatic in-plane C–H deformation vibration, respectively (Fig. 35). The vibrational band at 1233 cm^{-1} can be attributed to the hydroquinone molecule, an aromatic compound formed by *p*-BQ solution in water via a protonation reaction (Bratescu et al. 2011b).

The increase of the CARS signal strength can be explained by the effect of the electric field of the plasma, which determines the polarization of the *p*-BQ molecules and the orientation of the molecules with the electric dipole arranged along the electric field. The enhancement of the CARS signal intensity occurred when the pulsed electric field of the plasma was applied synchronously with the laser pulse to the *p*-BQ solution in water. This effect indicates that the *p*-BQ molecules are polarized and oriented under the external electric field of plasma. It was found that the enhancement of the CARS signal intensity of the *p*-BQ vibrational bands decays after switching off the HV pulse of the plasma, with a time constant of about 20 μs.

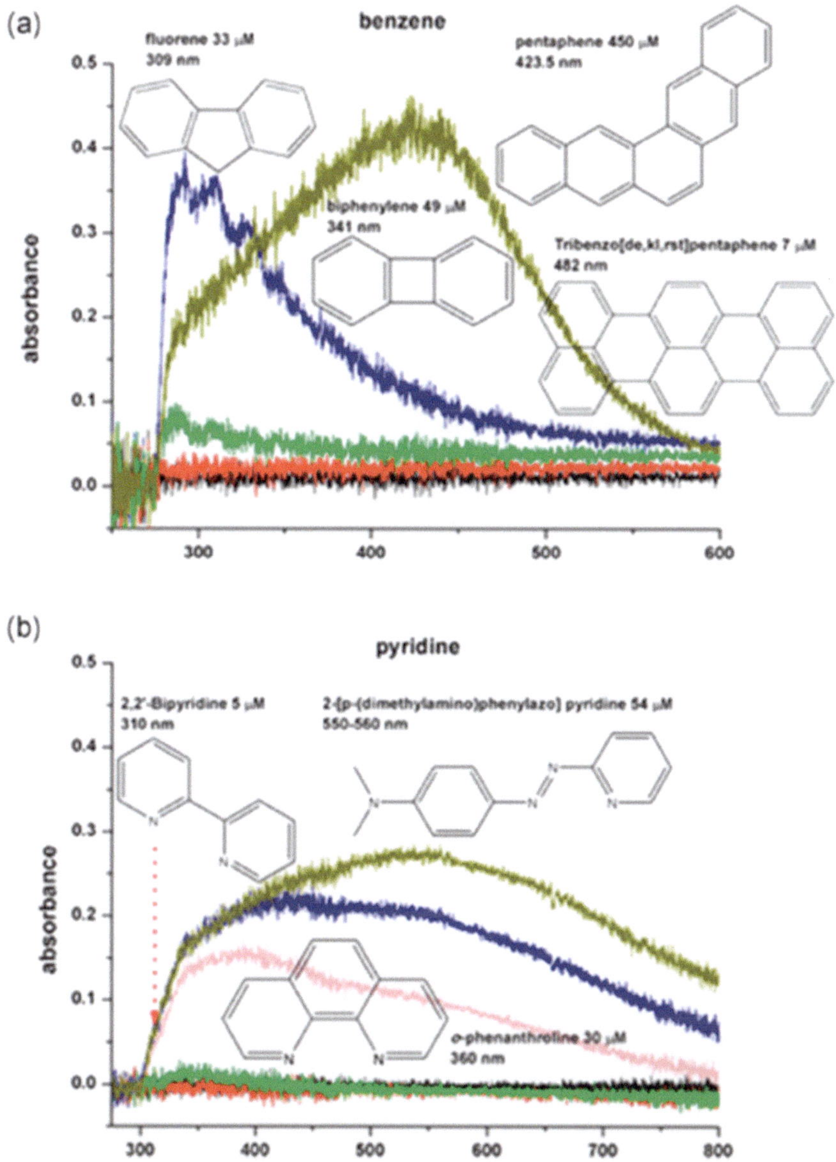

Fig. 33 In-situ BAS-sol results in **a** benzene, **b** pyridine, and **c** aniline. New low molecular weight chemical compounds were identified during the solution plasma operation. From the absorbance, knowing ε, the molar concentration was determined (reprinted from Bratescu et al. 2019 Copyright, with permission from Wiley Publishing)

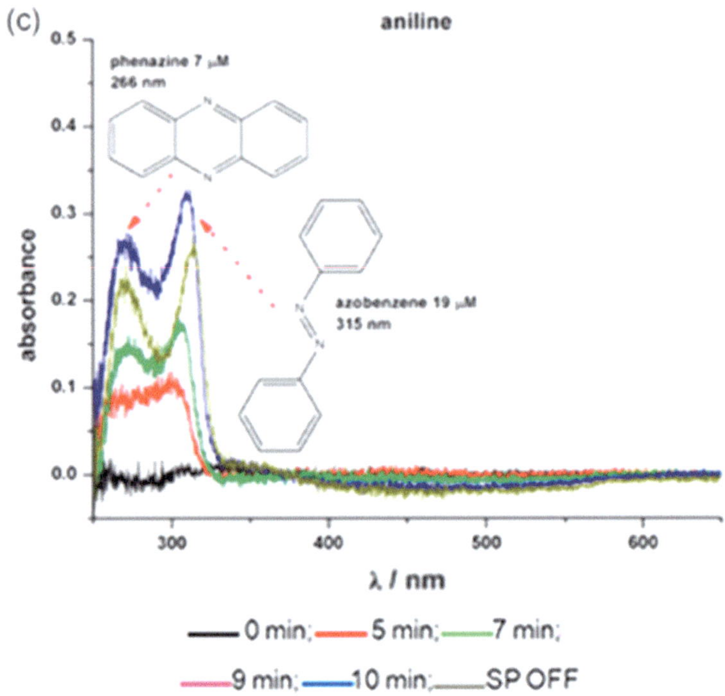

Fig. 33 (continued)

2.9.5 Electrical Probe Measurements

Solution plasma in an aqueous solution of sodium chlorate was investigated by measuring the floating potentials in various conditions (Chen et al. 2013). The potential values of the solution plasma in the gas and liquid phases are measured with an electrostatic probe. The time evolution and the radial profiles of the floating potential for the solution phase reveal that the dominant negative charges in the gas phase are the species of negative ions, such as OH^- and O_2^-., rather than electrons. A positive potential gradient at the interface between gas and solution phases exists due to an electric double layer caused by the separation of low-mass positive ions of H^+. and high-mass negative ions of OH^- and s O_2^- nearest to the interface (Fig. 36).

A modified electrical probe coupled to a low pass filter measured plasma potential in SP in different regions, including the interface between plasma and solution (Niu et al. 2022). SP was generated in benzene, toluene, phenol, and aniline. The interface between the plasma and solution was assumed to be like a semiconductor material with the Fermi level at the value of the plasma potential. The electron energy was mainly transferred to molecules with lower Fermi levels at the interface between plasma and solution. When the molecule had a lower ionization energy, its ionization could easily occur. Further analyzing the structural characteristics of carbon products,

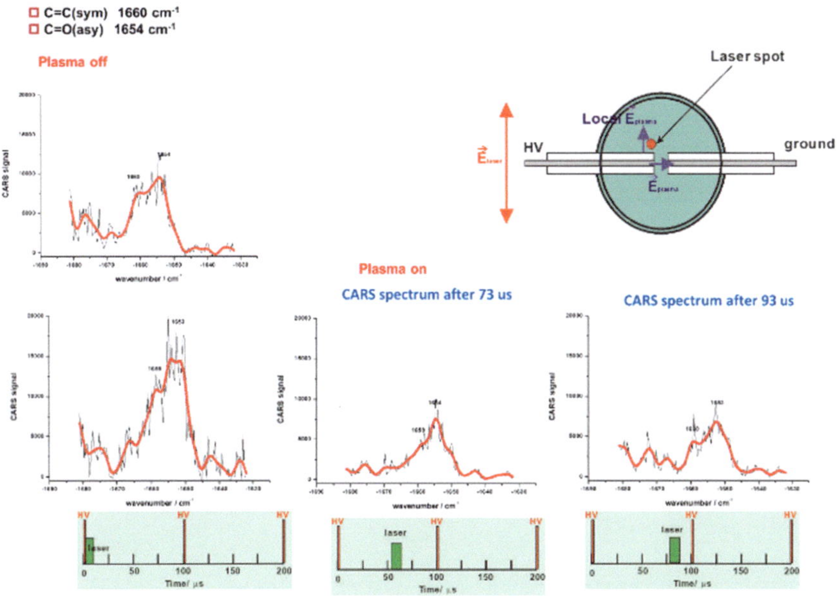

Fig. 34 The influence of the local electric field of plasma on the CARS signal intensity of the vibrations of *p*-BQ molecules in a 20 mM aqueous solution corresponding to the C=C symmetrical vibration at 1660 cm^{-1} and the C=O asymmetrical vibration at 1654 cm^{-1} in different conditions: plasma off; plasma on and the delay time between the plasma and the laser firing pulses of 0 µs, 73 µs, and 93 µs

it was found that the temperature difference at the interface between the plasma and solution determines the structure of the carbon material.

2.10 Collisions Processes in Solution Plasma

In the previous chapter, we have shown that the species from plasma are produced through reactions involving electrons, ions, and neutrals in the plasma gas phase (Tables 2). The reactions on a surface exposed to plasma and the generation of other species from the electrode surface. It was also shown to what extent these reactions can occur in SP.

In an elastic collision, the electron energy is kept constant. As was shown during the breakdown, the electron gains energy due to the external electric field. When an electron collides with a molecule or atom, the kinetic energy of electrons is transformed into potential energy. The molecule or atom is ionized or excited, depending on electron energy. Collision ionization of the molecule or atom in a strong electric field is the essential process that causes the gas breakdown. Ionization due to photons represents the interaction of radiation with matter. Photoionization occurs when the

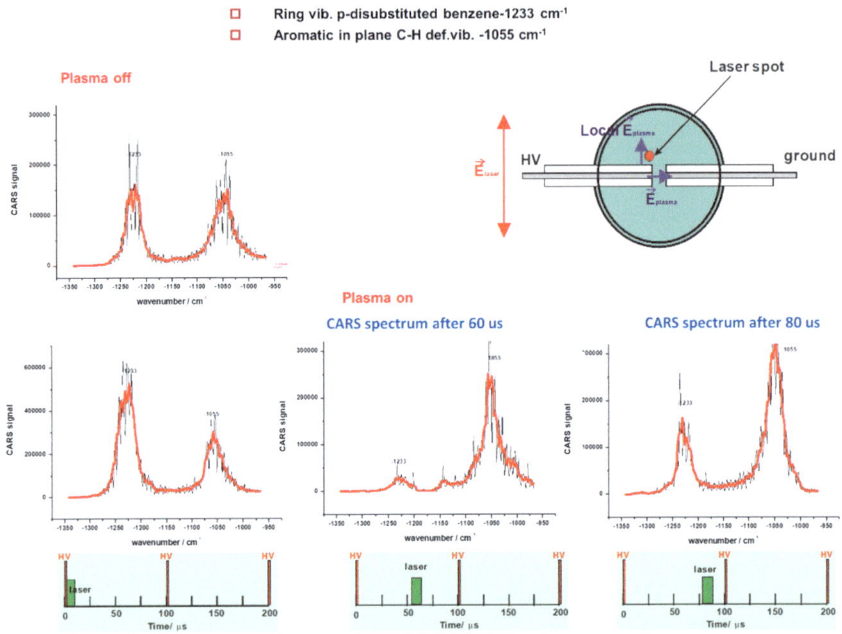

Fig. 35 The influence of the local electric field of plasma on the CARS signal intensity of the vibrations of *p*-BQ molecules in a 20 mM aqueous solution corresponding to the *p* substituted benzene at 1233 cm^{-1} and the aromatic in-plane CH deformation vibration at 1055 cm^{-1} in case of plasma off; plasma on, and the delay time between the plasma and the laser firing pulses of 0 µs, 60 µs, and 80 µs

energy of photons absorbed by an atom or molecule exceeds the ionization energy.

$$A + h\nu \rightarrow A^{+} + e^{-},$$

where A represents an atom or a molecule and $h\nu$ the energy of the photon.

Photoionization is a secondary ionization process. This process is essential in the mechanism of the streamer breakdown and the corona discharges. If the photon energy is less than the ionization energy, this energy produces the excitation of the atom or molecule at different energy levels, which means photoexcitation. Ionization by metastable interactions comes into operation long after excitation, and it has been shown that these reactions are responsible for long-time lags observed in some gases. The de-excitation time of some excited atoms or molecules on different electronic states can be even seconds. The names of these excited states are metastable states, and these atoms or molecules are called metastables. Metastables have a relatively high potential energy that can produce the ionization of neutral atoms or molecules. The ionization reaction by collision with a metastable is

$$A + B_{m} \rightarrow A^{+} + B + e^{-},$$

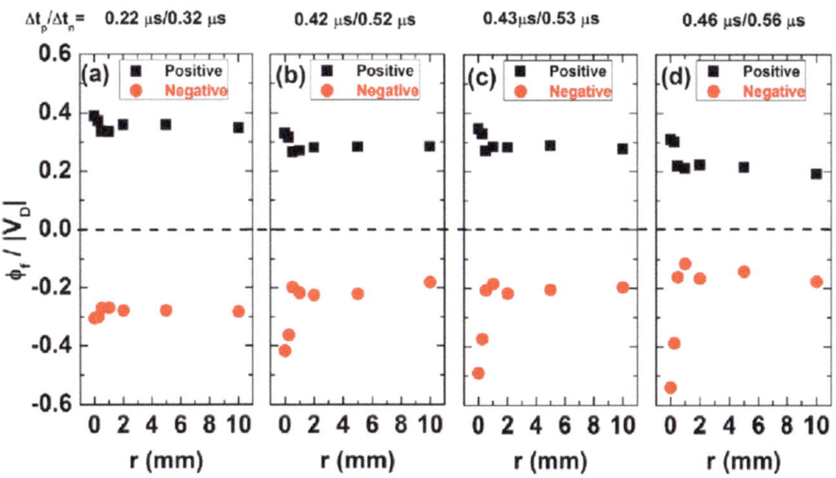

Fig. 36 The distribution of normalized floating potentials $\left(\Phi_f/|V_D|\right)$ as a function of time for $\frac{\Delta t_p}{\Delta t_n}$ =. **a** 0.22/0.32, **b** 0.42/0.52, **c** 0.43/0.53, and **d** 0.46/0.56. A solid circle and solid cube represent the $\left(\Phi_f/|V_D|\right)$ of the negative and positive pulse periods, respectively. Δt_p and Δt_n are the elapsed time counted from 0 μs for the positive pulse discharge plasma and 25 μs for the negative pulse discharge plasma. The voltage V_D is the voltage between the electrodes, and r is the distance from the center to the probe (reprinted from Chen et al. 2013 Copyright, with permission from AIP Publishing)

where A is the atom or molecule, and B_m is the atom or molecule in a metastable state.

The physical and chemical properties of the solution and electrodes determine the solution plasma parameters.

References

Banno M, Kanno K, Someya Y, Yui H (2015) Nanosecond time-resolved microscopic spectroscopy for diagnostics of an atmospheric-pressure discharge plasma formed in aqueous solution. Jpn J Appl Phys 54(6):066101

Baroch P, Anita V, Saito N, Takai O (2008) Bipolar pulsed electrical discharge for decomposition of organic compounds in water. J Electrostat 66(5–6):294–299

Baroch P, Potocký Š, Saito N (2011) Generation of plasmas in water: utilization of a high-frequency, low-voltage bipolar pulse power supply with impedance control. Plasma Sources Sci Technol 20(3):034017

Braithwaite NSJ (2000) Introduction to gas discharges. Plasma Sources Sci Technol 9(4):517–527

Bratescu MA, Saito N, Takai O (2011a) Redox reactions in liquid plasma during iron oxide and oxide-hydroxide nanoparticles synthesis. Curr Appl Phys 11(3):S30–S34

Bratescu MA, Cho SP, Takai O, Saito N (2011b) Size-controlled gold nanoparticles synthesized in solution plasma. J Phys Chem C 115(48):24569–24576

Bratescu MA, Hieda J, Umemura T, Saito N, Takai O (2011c) Analysis of organic pollutant degradation in pulsed plasma by coherent anti-Stokes Raman spectroscopy. J Vac Sci Technol, A 29(3):031302
Bratescu MA, Kim K, Saito N (2019) Quantitative spectrochemical analysis of solution plasma in aromatic molecules. Plasma Processes Polym 16(6):1–13
Bruggeman P, Schram D, González MÁ, Rego R, Kong MG, Leys C (2009) Characterization of a direct DC-excited discharge in water by optical emission spectroscopy. Plasma Sources Sci Technol 18(2):025017
Bruggeman P, Verreycken T, González MÁ, Walsh JL, Kong MG, Leys C, Schram DC (2010) Optical emission spectroscopy as a diagnostic for plasmas in liquids: opportunities and pitfalls. J Phys D Appl Phys 43(12):124005
Ceccato PH, Guaitella O, Le Gloahec MR, Rousseau A (2010) Time-resolved nanosecond imaging of the propagation of a corona-like plasma discharge in water at positive applied voltage polarity. J Phys D Appl Phys 43(17):175202
Chen Q, Kaneko T, Matsuda N, Hatakeyama R (2013) Potential structure of discharge plasma inside liquid directly measured by an electrostatic probe. Appl Phys Lett 102(24):244105
Clements RM, Smy PR (1973) Langmuir probe measurement of electron temperature in high-pressure plasmas. J Appl Phys 44(8):3550–3556
Clements RM, Smy PR (1974) The floating potential of a Langmuir probe in a high-pressure plasma. J Phys D Appl Phys 7(4):551–562
Demtröder W (2008) Laser spectroscopy. volume 1: basic principles, 4th edn. Springer, Berlin Heidelberg, Chap 3, pp 61–97
Gigosos MA, González MA, Cardenoso V (2003) Computer simulated Balmer-alpha, -beta and -gamma Stark line profiles for non-equilibrium plasmas diagnostics. Spectrochim Acta, Part B 58(9):1489–1504
Goyette AN, Lawler JE, Anderson LW, Gruen DM, McCauley TG, Zhou D, Krauss AR (1998) Spectroscopic determination of carbon dimer densities in Ar–H_2–CH_4 and Ar–H_2–C_{60} plasmas. J Phys D Appl Phys 31(15):1975–1986
Fantz U (2006) Basics of plasma spectroscopy. Plasma Sources Sci Technol 15(4):S137–S147
Heo YK, Lee SH, Bratescu MA, Kim SM, Lee GJ (2017) Generation of non-equilibrium condition in solution plasma discharge using low-pass filter circuit. Plasma Processes Polym 14(5):1600163
Hirayama K (1967) Handbook of ultraviolet and visible absorption spectra of organic compounds. Plenum Press Data Division, New York, Fuji Photo Film Research Laboratories, Odawara, Japan
Jameson EC (2001) Description and development of electrical discharge machining. Society of Manufacturing Engineers, Dearborn, Michigan, USA
Jiang LY, Dziedzic P, Spacil Z, Zhao G-L, Nilsson L, Ilag LL, Córdova A (2014) Abiotic synthesis of amino acids and self-crystallization under prebiotic conditions. Sci Rep 4:6769
Kattan R, Denat A, Lesaint O (1989) Generation, growth, and collapse of vapor bubbles in hydrocarbon liquids under a high divergent electric field. J Appl Phys 66(9):4062–4066
Kattan R, Denat A, Bonifaci N (1991) Formation of vapor bubbles in nonpolar liquids initiated by current pulses. IEEE Trans Dielectr Electr Insul 26(4):656–662
Laux CO, Spence TG, Kruger CH, Zare RN (2003) Optical diagnostics of atmospheric pressure air plasma. Plasma Sources Sci Technol 12(2):125–138
Lewis TJ (1988) An overview of electrical processes leading to dielectric breakdown in liquids. In: Kundhart EE et al (eds) The liquid state and its electrical properties. Plenum, New York, pp 431–443
Lewis TJ (1994) Basic electrical processes in dielectric liquids. IEEE Trans Dielectr Electr Insul 1(4):630–643
Little PF (1956) Secondary effects. In: Flügge S (ed) Handbuch der physic, vol 21, pp 574-663
Locke BR, Sato M, Sunka P, Hoffmann MR, Chang J-S (2006) Electrohydraulic discharge and nonthermal plasma for water treatment. Ind Eng Chem Res 45(3):882–905
Loeb LB, Meek JM (1940a) The mechanism of spark discharge in air at atmospheric pressure I. J Appl Phys 11(6):438–447

Loeb LB, Meek JM (1940b) The mechanism of spark discharge in air at atmospheric pressure II. J Appl Phys 11(7):459–474

Luque J, Hudson EA, Booth JP (2003) CF $A^2\Sigma^+$–$X^2\Pi$ and $B^2\Delta$–$X^2\Pi$ study by broadband absorption spectroscopy in a plasma etch reactor: determination of transition probabilities, CF $X^2\Pi$ concentrations, and gas temperatures. J Chem Phys 118(2):622–632

Miller SL (1953) A production of amino acids under possible primitive earth conditions. Science 117(3046):528–529

Miller SL (1955) Production of some organic compounds under possible primitive Earth conditions. J Am Chem Soc 77(9):2351–2361

Miron C, Bratescu MA, Saito N, Takai O (2010) Time-resolved optical emission spectroscopy in water electrical discharges. Plasma Chem Plasma Process 30(5):619–631

Miron C, Bratescu MA, Hieda J, Saito N, Takai O (2011) Optical diagnostics of bipolar electrical discharges in HCl, KCl, and KOH solutions. J Appl Phys 109(12):123301

Morishita T, Ueno T, Panomsuwan G, Hieda J, Yoshida A, Bratescu MA, Saito N (2016) Fastest formation routes of nanocarbons in solution plasma processes. Sci Rep 6:36880

Murphy EJ (1970) The generation of gases in the dielectric breakdown of oil. J Phys D Appl Phys 3(7):917–921

Niu J, Chokradjaroen C, Sawada Y, Wang X, Saiti N (2022) Plasma-solution junction for the formation of carbon material. Coatings 12:1607

NIST atomic spectra database (n.d.). Retrieved from http://physics.nist.gov/PhysRefData/

Pearse RWB, Gaydon AG (1950) The identification of molecular spectra. Wiley, New York

Raizer YP (1987) Gas discharge physics. Springer, Berlin Heidelberg, New York, Tokyo, Hong Kong, Barcelona, Budapest Chap 12

Saito N, Bratescu MA, Hashimi K (2018) Solution plasma: a new reaction field for nanomaterials synthesis. Jpn J Appl Phys 57(1):0102A4

Saito N, Ueno T, Bratescu MA, Hieda J (2019) Synthesis of nanomaterials using solution plasma process. In: Setsuhara Y et al (eds) Novel structured metallic and inorganic materials. Springer, Chap 24, pp 343–355

Schönland BFJ (1956) The lightning discharge. In: Flügge S (ed) Handbuch der physic, vol 22, pp 576–628

Takai O (2008) Solution plasma processing (SPP). Pure Appl Chem 80(9):2003–2011

Takai O, Bratescu MA, Ueno T, Saito N (2013) Synthesis of nanomaterials by solution plasma processing. In: Fukuda T et al (eds) Micro-nano mechatronics. IntechOpen, Chap 6, pp 109–122

Townsend JS (1910) Theory of ionization of gases by collision. Constable & Company Ltd., London

Ushakov VY. Klimkin VF, Korobeynikov SM (2007) Impulse breakdown of liquids power systems. Springer, Berlin, Chap 7.1

Ushakov VY, Klimkin VF, Korobeynikov SM (2007b) Impulse breakdown of liquids power systems. Springer, Berlin Chap 1

Wadhwa CL (2020) High voltage engineering. New Age International Publishers, Chapter 1

Wilsing J (1899) On the interpretation of the typical spectrum of the new stars. Astrophys J 10(2):113-125

Yang Y, Cho YI, Fridman A (2012) Plasma discharge in liquid. CRC Press, Chap 3

Yubero C, Calzada MD, García MC (2005) Using the Stark broadening of the Hα, Hβ, and Hγ lines for the measurement of electron density and temperature in a plasma at atmospheric pressure. J Phys Soc Jpn 74(9):2249–2254

Electrochemistry in Solution Plasma

Nagahiro Saito and Chayanaphat Chokradjaroen

Abstract In solution plasma (SP), the collision of particles (e.g., charged particles, neutral molecules), electron exchange, sputtering of metal surface, vaporization, and generation of ultra-violet (UV) are generally occurred. These phenomena usually result in the formation of reactive species and facilitate chemical reactions. To elucidate, since the components of SP are composed of plasma in gas phase surrounded by liquid phase, there are two reactions routs, previously proposed, including through plasma area where the collision of electrons and ions intensively occurred and through plasma-liquid interface where the chemical reactions are considered to be a heterogeneous reaction. In this chapter, the explanation relating to the reactions in each rout in different solutions, such as aqueous solutions and organic solutions, is provided. Moreover, the synthesis of metal nanoparticles via the SP through different techniques, including reduction of precursor solution and sputtering of metal nanoparticle from electrode surface, is thoroughly described.

1 Introduction

Reactions in nonthermal plasma at atmospheric pressure have been studied for over a century. These studies have simulated plasma application in water purification, nanomaterials synthesis, and catalyst-free reactions (Kaneko et al. 2017; Mai-Prochnow et al. 2021; Saito et al. 2018). These applications utilize highly reactive species, including electrons, radicals, atomic and molecular ions, and photons, which are products of the plasma systems that either by suspending the plasma electrode over the liquid surface (Fig. 1a) or by submerging the electrode into the liquid (Fig. 1b, c)

N. Saito (✉)
Department of Chemical Systems Engineering, Graduate School of Engineering, Institute for Innovation for Future Society, Nagoya University, Furocho, Chikusa, Nagoya, Japan
e-mail: saito.nagahiro.z7@f.mail.nagoya-u.ac.jp

C. Chokradjaroen
Institute for Aqua Regeneration, Shinshu University, Matsumoto, Nagano, Japan
e-mail: chokradjaroen.c@shinshu-u.ac.jp

N. Saito (ed.), *Solution Plasma Chemistry*, Lecture Notes in Chemistry 115,
https://doi.org/10.1007/978-981-95-0966-9_4

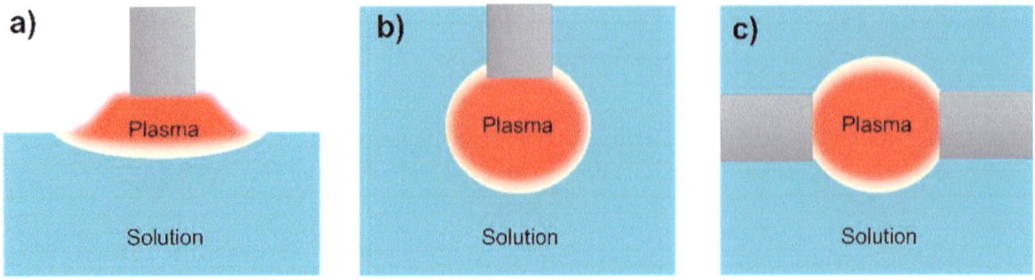

Fig. 1 Examples of plasma-liquid system schematics; **a** plasma generated on surface of a liquid; **b** plasma generated in the liquid by submerged one electrode; **c** plasma generated in the liquid by submerged two electrodes, refering to the conventional solution plasma system

(Elg et al. 2021). Both the plasma zone and plasma-liquid interface play an important role as reaction fields in the solution plasma.

As mentioned in the previous chapter, plasma is conductive in the same manner as metals, but particles in plasma, including electrons and positively and negatively charged ions, can occupy and move in the space between electrodes. For the plasma in liquids, like solution plasma, the phenomenon occurs between two electrodes placed in solution, which is analogous to an electrochemical cell. However, not only metal electrodes but also liquid surfaces can act as electrodes where electrons can be gained and donated. The plasma-liquid interface plays an important role as a reaction field in the solution plasma. In an aspect of the reaction area, the plasma-liquid interface area is larger and outside the plasma. Therefore, the reactions dominantly occur at the plasma-liquid interface through the transfer of electrons and ions (Saito et al. 2018). The complex conditions in the solution plasma, consisting of ultraviolet radiation, radicals, electrons, and ions, results in non-Faradaic processes. Accordingly, the charge transfer to drive reactions is still not clear and needs to be identified. Especially, electrons are the origin of the plasma-induced reaction. Thus, the understanding of electron transfer plays a significant role in the utilization of plasma for several applications.

This chapter points out that the electron transfer can relate to the reduction–oxidation reaction, which may be a key for plasma-liquid interface reactions (Elg et al. 2021). The concept of electrochemistry will be applied to explain phenomena in the solution plasma and provide a clear understanding of reduction–oxidation reactions at the plasma-liquid interface of the solution plasma. Some previous studies will be described and discussed.

2 Electrochemistry in Solution Plasma

In solution plasma systems, electrochemical equilibrium theory is crucial for explaining the interactions between plasma and liquid. This theory focuses on two main aspects: (1) surface reactions at the electrode and (2) reactions involving the solvent or solute. Both aspects are closely linked to the standard electrode potential (E°) and changes in Gibbs free energy (ΔG), which determine the spontaneity and direction of redox reactions. Furthermore, the pH dependence of electrode potential also plays a significant role in influencing the progression of redox reactions.

(1) Surface Reactions at the Electrode: Changes in the material on the electrode surface affect its ability to emit electrons, leading to changes in the plasma state.
(2) Reactions of the Solvent or Solute: These reactions are governed by the standard electrode potential relative to the plasma floating potential (or molecular energy levels such as HOMO, LUMO, and Fermi level), which determine whether a molecule is oxidized or reduced.

Below, we will delve into these aspects in detail, emphasizing the relationship between electrode potential, free energy, and pH dependence.

2.1 *Surface Reactions at the Electrode*

In solution plasma systems, surface reactions at the electrode involve electron transfer between the electrode and the plasma, which affects the plasma's electron density and charge distribution. These reactions can be understood by considering the relationship between the standard electrode potential and the Gibbs free energy change.

2.1.1 Standard Electrode Potential and Gibbs Free Energy

The change in Gibbs free energy (ΔG) during an electrochemical reaction is related to the electrode potential (E) by the following equation:

$$\Delta G = -nFE \tag{1}$$

where:

ΔG is the Gibbs free energy change (J/mol),
n is the number of electrons involved in the reaction,
F is the Faraday constant (96,485 C/mol),
E is the electrode potential (V).

A negative ΔG indicates a spontaneous reaction, while a positive ΔG suggests a non-spontaneous reaction. A positive electrode potential leads to a negative ΔG, promoting the reaction (oxidation or reduction), whereas a negative electrode potential makes the reaction less favorable.

2.1.2 pH Dependence of Electrode Potential

The electrode potential depends on the pH of the solution, which influences the redox reactions occurring at the electrode. This relationship is described by the Nernst equation:

$$E = E^0 - 0.059n\log\left(\frac{[Ox]}{[\mathrm{Re}d]}\right) - \frac{0.059}{n}\cdot \mathrm{pH}$$

where:

E^0 is the standard electrode potential,
$[Ox]$ is the concentration of the oxidized species,
$[Red]$ is the concentration of the reduced species,
pH represents the hydrogen ion concentration in the solution,
In the Nesnst equation, the temperature is 298 K.

As the pH increases (becomes more basic), the electrode potential decreases, favoring reduction reactions and making oxidation reactions less likely. Conversely, at lower pH (acidic conditions), oxidation reactions are more favorable.

This dependence on pH can be visualized in the potential-pH diagram (Pourbaix diagram), which plots the stability regions for different species (metal, ions, oxides) as a function of pH and electrode potential. The diagram shows three primary regions:

- Immunity: The metal remains stable and unreactive.
- Corrosion: The metal dissolves, forming ions.
- Passivation: A protective oxide layer forms on the metal surface.

This diagram is a valuable tool for predicting how changes in pH and potential will affect the redox behavior of an electrode in a solution plasma system.

2.1.3 Surface Changes and Plasma State

The Pourbaix diagram is a graphic representation of the thermodynamic stability of different species in an aqueous system. By using the Pourbaix diagram, one can predict whether the electrode surface is prone to oxidation or reduction under certain pH and potential conditions. As the surface composition changes, such as when the electrode forms a protective oxide layer, the conductivity and electron emission capability of the electrode change significantly, influencing the overall plasma state, including its electron density and temperature. The non-LTE condition corresponding to the plasma gas phase might change, too.

2.2 *Reactions of the Solvent or Solute*

In solution plasma systems, the redox reactions of solutes or solvents are controlled by the interaction between the standard electrode potential (E^0) and the plasma floating potential (V_f) defined in Eq. 23 in Chap. "Characteristics of Solution Plasma". This interaction determines whether a molecule is reduced or oxidized and is closely related to changes in Gibbs free energy (ΔG).

2.2.1 Standard Electrode Potential and Molecular Energy Levels

A standard electrode potential (E^0) of molecules reflects its tendency to accept or donate electrons during a redox reaction. This potential is related to the Gibbs free energy change. At the plasma-liquid interface, the redox behavior depends on whether the plasma floating potential (V_f) is higher or lower than the standard electrode potential of the solute. If V_f is higher than the solute's reduction potential, the plasma donates electrons, causing a reduction reaction. A more positive potential (higher than the standard electrode potential) results in a negative ΔG, making the reduction reaction spontaneous:

$$A^{n+} + ne^- \rightarrow A$$

If V_f is lower than the solute's oxidation potential, the plasma accepts electrons, promoting an oxidation reaction. A more negative potential (lower than the oxidation potential) also leads to a negative ΔG, facilitating spontaneous oxidation:

$$A \rightarrow A^{n+} + ne^-$$

2.2.2 Molecular Energy Levels (HOMO and LUMO)

Redox behavior can also be understood in terms of molecular energy levels, specifically the HOMO (Highest Occupied Molecular Orbital) and LUMO (Lowest Unoccupied Molecular Orbital) levels:

- Reduction: Molecules with a low LUMO level are more likely to accept electrons and be reduced by plasma.
- Oxidation: Molecules with a high HOMO level are more likely to lose electrons and be oxidized.

2.2.3 PH and Redox Reactions: Role of the Pourbaix Diagram

The pH of the solution has a significant impact on the redox behavior of solutes and solvents. In acidic conditions (low pH), reduction reactions are more favorable due to the abundance of protons (H^+), while in basic conditions (high pH), oxidation reactions are promoted due to the presence of hydroxide ions (OH^-).

The Pourbaix diagram can be used to predict how pH affects the redox behavior of a given solute or solvent. By showing the stability regions of different species (ions, neutral molecules, oxides) as a function of pH and potential, the diagram helps determine under which conditions reduction or oxidation will dominate.

3 Reduction and Oxidation at Plasma-Liquid Interface

The electron transfer process between plasma and liquid can be in the same manner as that between solid electrodes and electrolytes in an electrochemical cell to promote reduction–oxidation (redox) reactions. In other words, the electron exchange at the plasma-liquid interface can sustain plasma and reactions, such as ionization, inside a plasma, as shown in Fig. 2, leading to the formation of reactive species. These reactive species can further move through the plasma-liquid interface and result in chemical reactions at or near the interface. Much research has been proposed to study aiming to understand the plasma-liquid interface reactions. Richmond and co-workers demonstrated that a solid cathodic electrode could be replaced by an atmospheric pressure micro-plasma formed in an Ar gas flow to the electrolyte surface, acting as a gaseous cathodic electrode (Richmonds et al. 2011). Several parameters were found to be involved, such as discharge parameters (e.g., discharge current), electron density (flux) and distribution, electron energy, and solution potential. Moreover, Shirafuji and co-workers also summarized the concept of electrochemistry in a plasma-liquid interface, as shown in Fig. 3 (Shirafuji et al. 2013).

Electrical discharge in liquid, like solution plasma, shows three main mechanisms. One is the electric breakdown, which generally relates to an electron avalanche, which was previously described in Chap. “Characteristics of Solution Plasma”. The breakdown in liquid is usually suggested to be the phase change, i.e., liquids phase to gas phase, due to the natural formation of bubbles caused by electrolysis. In other words, the plasma is in the gas phase and surrounded by the liquid phase. After the breakdown, the propagation and maintenance of plasma are procedures. The maintenance of plasma has been reported to require secondary electrons. Secondary electrons are electrons generated by other radiation via the collision of ions or photons with sufficiently high energy, i.e., exceeding the ionization potential, to neutral atoms and molecules, such as on the surface of metal electrodes. These phenomena occur in the same manner for discharges in both gas and liquid, as shown in Figs. 4 and 5.

In the plasma, only vaporization and liquefaction processes generally occur, and there is a difference in a local electric field and a local density gradient. At the interfaces, for either plasma-liquid or plasma-solid, sheathing is a phenomenon in

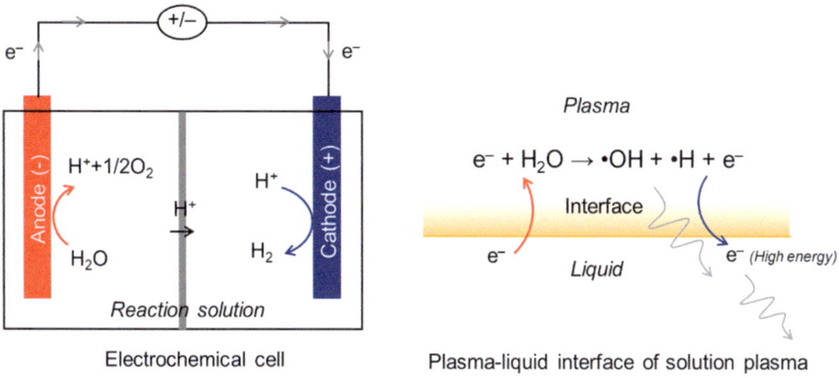

Fig. 2 Comparison between reduction–oxidation reactions in conventional electrochemical cell and the solution plasma system

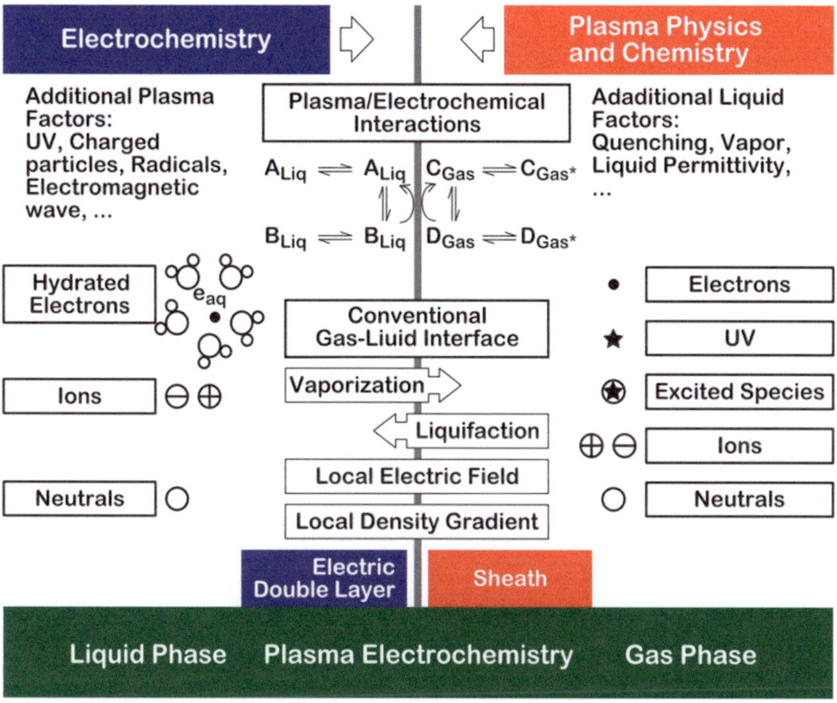

Fig. 3 The concept of electrochemistry in plasma-liquid interface. Reprinted with permission from Shirafuji et al. (2013)

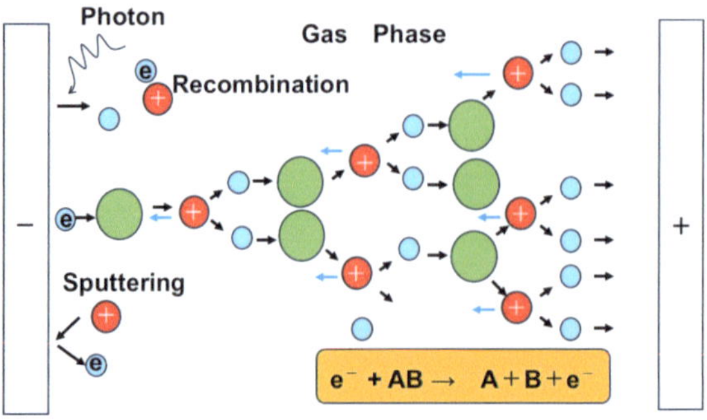

Fig. 4 Illustrating of electrical discharge in gas phase, showing electron collision and transfer

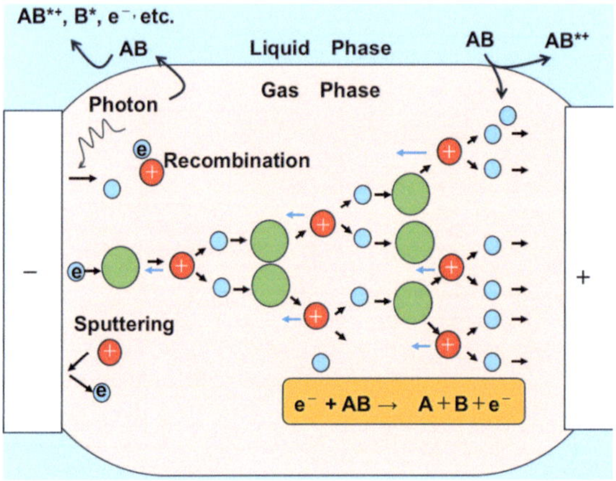

Fig. 5 Illustrating of electrical discharge in the gas phase, showing electron collision and transfer

that the charges accumulate to shield the potential of the plasma with the contacting body is called the plasma sheath. In a gas-phase discharge, when a plasma comes into contact with any solid material, an accumulation of charges is formed on the contact surface due to the effect of Debye shielding or electric field screening. In the liquid-phase discharge, ions play an important role in the plasma. Thus, in an electron-rich environment, the ions with different electronegativity can result in the formation of heavy particles. Chen and a co-worker reported that the electrical discharge in water usually produced $^{\bullet}O$ and $^{\bullet}OH$ and $^{\bullet}O$, then may combine to form O_2. OH^- and O_2^- are easily formed in water vapor by electron attachment (Chen et al. 2013). In addition, the OH^- in water can be in the hydrated form, i.e., $(H_2O)_mOH$~ (m is the number of hydrated water molecules). Therefore, the plasma consists of ions, including H^+, O_2~, and $(H_2O)_mOH^-$, and electrons. The electrons in the plasma easily undergo attachment because it is a reaction with relatively low threshold energy (Itikawa and Mason 2005; Joshi and Thagard 2013). As a result, the inside of plasma is dominated by negative ions. The masses of the negative species, e.g., O_2^- and $(H_2O)_mOH^-$, are much larger than the mass of the positive species, e.g., H^+. Accordingly, the negatively charged ions usually tend to remain in the plasma area while the positively charged ions move toward and locate near the plasma-liquid interface. This configuration of the charged species near the plasma-liquid interface induces the electrical double layers in the plasma-liquid interface (Chen et al. 2013), as shown in Fig. 6. The bilayer structure is the source of kinetics in the chemical reaction process.

Electrical discharges in either gas or liquid, the contact between plasma and solid, are a primary position where the secondary electrons can be produced through the collision of ions and photons to sustain the plasma. On the other hand, the electrons can transfer through the contact of plasma and solution, which is the environment with high molecular density acting as a solid surface. Therefore, the contact between plasma and liquid becomes an important topic to discuss. Some previous research proposed that, in the plasma in contact with the liquid, the liquid itself (e.g., water) plays an important part in plasma generation. When the plasma is positively biased, positive ions move to the solution surface and cause water ionization (Fig. 7a) (Elg et al. 2021; Goodman et al. 1973). It is similar to heavy ion radiolysis. When the negatively biased plasma is generated, relatively low-energy free electrons are injected from the gaseous plasma into water (Fig. 7b). Compared to conventional electrolysis, where electron transfer occurs at the solid surface of an electrode with adsorbed surface species, in liquid, electrons become solvated and are presently free of any solid surface within the volume of water, regardless of the polarity of the plasma (Elg et al. 2021). These phenomena can also be used to explain the chemistry at the plasma-liquid interface in the solution plasma. The liquid may act like a solid surface where the electrons and ions can transfer. However, the liquid surface can allow electrons and ions freely move and participate in the chemical reaction.

To fundamentally explain, by being a part of the plasma electrical circuit, the liquid must facilitate charge transfer across the plasma-liquid interface for the plasma to form and be sustained. For example, Richmonds and co-workers (Richmonds et al. 2011) demonstrated the electron transfer at the plasma-liquid interface through

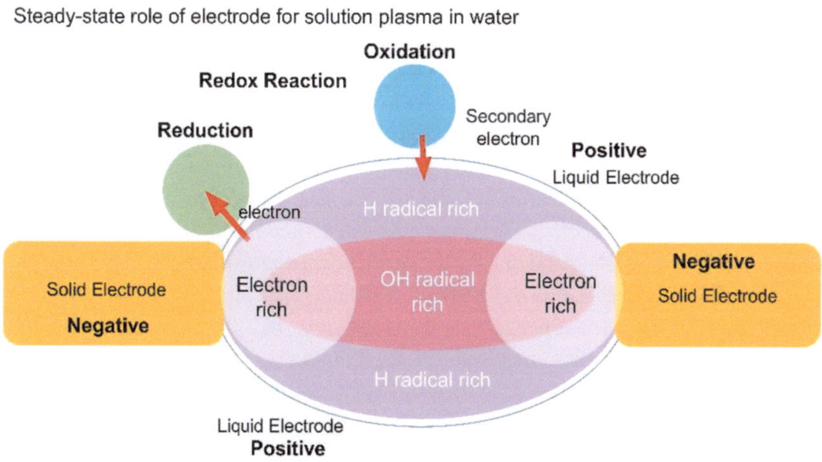

Fig. 6 Electrical double layer structural diagram of plasma in a liquid phase

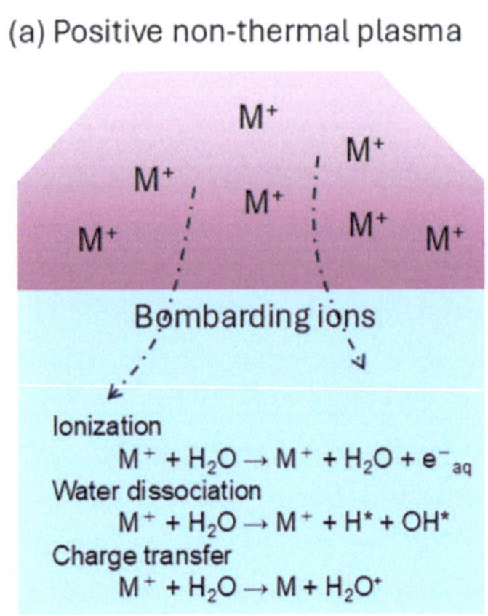

Fig. 7 Schematics of a plasma-water interface where the plasma is electrically **a** positive and **b** negative relative to the water such that free plasma electrons are directed toward the water surface. Reprinted with permission from Elg et al. (2021)

the reduction of ferricyanide ($[Fe(CN)_6]^{3-}$) to ferrocyanide ($[Fe(CN)_6]^{4-}$), and the reduction rate is found to depend on the discharge current. The study exhibited that the electron transfer at the plasma-liquid interface could initiate and control electrochemical reactions, which is evidence of the electrochemistry approach via the plasma system. Moreover, Witzke and co-workers (2012) presented that plasma electrons are involved in electrolytic reactions, in the case of water, leading to the conversion of protons (H^+) to hydrogen gas. The result showed the occurrence of water electrolysis, which was indirectly characterized by pH measurements. The evolved hydrogen and oxygen gases were believed to react in the plasma, such that the charge transfer reactions and plasma gas reactions were inherently coupled.

Non-equilibrium plasma in the solution requires secondary electrons to sustain the plasma state because it lacks electrons. In general, secondary electrons are supplied from electrodes, generally metal electrodes. In the case of the solution plasma, the solution component also functions as electrodes (i.e., liquid electrodes), enabling the plasma to gain secondary electrons from the solution, i.e., molecules, as shown in Fig. 6. Upon the emission of the secondary electrons, radical cations form in the solution. In other words, electron transfer occurs all the time during the discharge.

The interaction between the plasma and the molecule in the solution plasma system was used to explain and clarify the reaction mechanism of the polymerization of carbon materials (Morishita et al. 2016). The electron can be withdrawn from the benzene molecules to form benzene cations radicals, like the oxidation process in conventional electrochemical reactions. After that, the benzene cations radicals gain electrons from plasma and hydrogen gas, a product of the recombination, which initiates the polymerization. The transfer of excited electrons from the plasma to the molecules in solution led to the linkage between six-membered ring structures to form a carbon framework. The process could be analogous to the reduction reaction.

Thus, it can imply that the molecules in the solution act like an anode and cathode in the system, while the plasma is an excitation field, drawing electrons out and returning excited electrons to the solution. Moreover, the charge transfer processes are also a strong interdependence of ions and radicals, and the electrons, withdrawn and injected by a plasma to and from important for driving solution-phase chemistry. From the perspective of electrochemical analysis, the reaction step in solution plasma reaction field, which mainly indicates the electron transfer process, has been focused on.

In general, the Fermi level of electrons in the plasma always lies above the vacuum energy level. This is because the plasma always appears with high potential when it contacts any object. In other words, the plasma always passes electrons outwards. Fermi energy (E_F) is the energy level that is occupied by the electron at a temperature equal to 0 K. The E_F of a system is defined as a function of temperature. The temperature of electrons in a metal is different from the temperature of electrons in a plasma. Electrons in metal are bounded within the metal surface, and hence, their Fermi level lies below the vacuum level; therefore, their energy potential (E_p) < 0. On the other hand, for laboratory plasmas, plasma potential (V_p) is the highest potential in a device containing plasma. In other words, plasma has the highest potential with respect to any potential exposure to it, which depends on the mass of reactive species and electron temperature. Hence, we can assume that the V_P is the highest energy available to electrons in the plasma, and the Fermi energy of electrons in plasma may be considered as $Ep \ll eV_p$ (Arumugam et al. 2020).

The plasma generated in a laboratory has been assumed to be a semiconducting material. It follows that the valence band of the plasma should be greater than the Fermi level present. Then, the conduction band of the plasma is always lower than the highest orbital, which can be filled by the electrons of any molecule with which it comes into contact. Based on this understanding, when a molecule encounters a plasma, acting as a semiconductor, the E_F level resonances occur at the contact surface because the system attempts to reach equilibrium. The resonance of the E_F levels leads to the transfer of electrons from the solution molecule with a higher HOMO to the conduction band of plasma in the solution plasma. It is noteworthy that this process coincides with the conventional γ-effect of secondary electron emission (Liu et al. 2018). In the γ process, the solvent molecule is excited, and the secondary electrons injected into the solution plasma are released as a result of the collision of the higher energy ions with the solvent molecule and the resulting energy transfer (Morishita et al. 2016).

In the solution plasma, the reaction steps include the initial step, excitation and reduction, and collision. The excitation of electrons in the solution plasma can be in the same manner as that of photochemical reactions. Once the electrons have been transferred to the valence band of the plasma, the first step of the initial step has been completed (Fig. 8). The second step is when the solvent molecule is in the excited state (Fig. 9). The electrons in the valence band of the plasma are excited by the high voltage electric field provided by the power supply to the conducting band of the plasma, where the electron possesses high energy enough to carry on further reactions (Fig. 10).

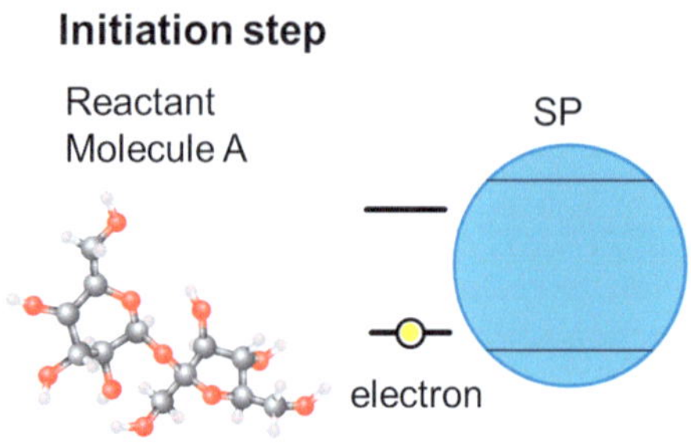

Fig. 8 Initial step of electron transfer between the solution plasma (SP) and the solution molecules

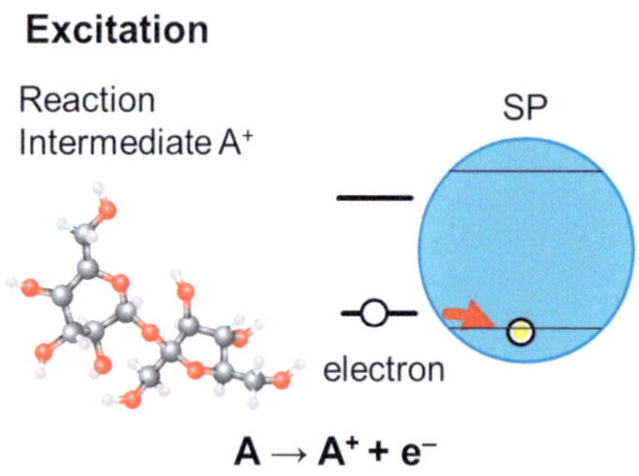

Fig. 9 Second step (excitation of solution molecule) of electron transfer between the solution plasma (SP) and the solution molecules

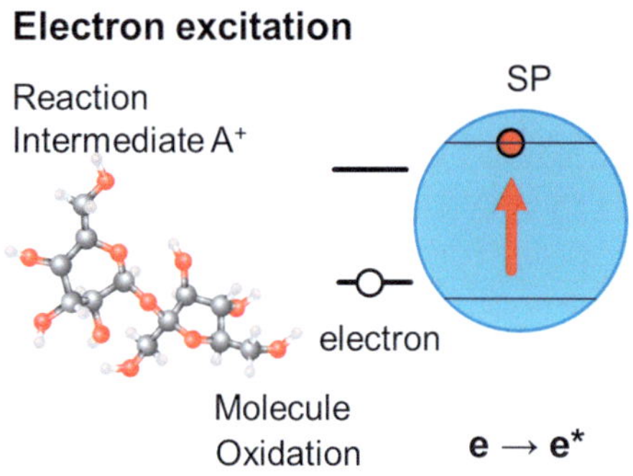

Fig. 10 Excitation of an electron in the solution plasma (SP)

The electrons are shortly present in the conducting band. Subsequently, they are quickly transferred to molecules with lower LUMO energy levels, thus achieving a reduction. At the same time, some of the electrons in the plasma gain further energy from the electric field and undergo a collision process with other molecules around them (Fig. 11). The processes are continuously proceeded, as prolonging the discharge.

Chae and co-workers (2017) proposed the fabrication of few-layer graphene from graphite in a 1-ethyl-3-methylimidazolium dicyanamide (EMIM-DCA) aqueous solution by the solution plasma. The formation of the few-layer graphene can be explained by the exchange of electrons via EMIM-DCA in water. The energy diagram

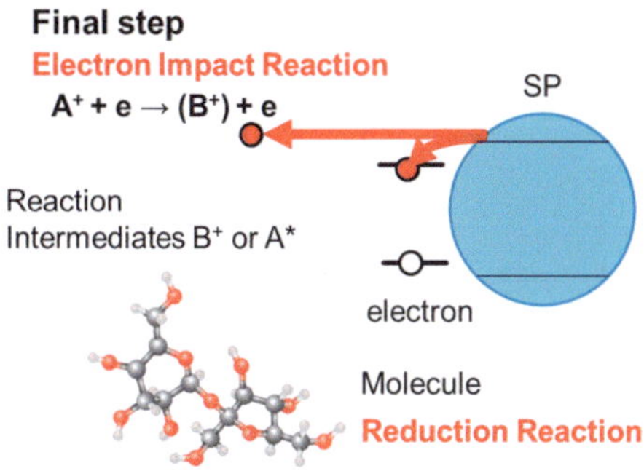

Fig. 11 Transfer and collision of high-energy electron to solution molecules in the solution plasma (SP)

of the solution plasma system in the EMIM-DCA aqueous solution with graphite flakes, corresponding to pH 7, is shown in Fig. 12. The energy scale (right side) is negative and relative to the vacuum level (VL). The energy scale on the left side is versus the normal hydrogen electrode (NHE). The work function of graphite was assigned to be −4.19 eV. The calculated ionization potential and electron affinity for the EMIM dimer in water at pH 7 were −4.47 eV and −3.49 eV, respectively. The corresponding values of ionization potential and electron affinity for DCA were −5.28 eV and −0.85 eV, respectively. The calculated values were consistent with those experimentally obtained from the CV measurements in 10 wt% EMIM-DCA aqueous solution. Thereby, the ionization potential and electron affinity for EMIM were found at −5.86 eV and −3.94 eV, respectively, and the corresponding values for DCA were determined to be −6.20 eV and −3.25 eV, respectively. The bonded states in solution plasma corresponded to the energy levels of the species present in plasma, such as atoms, molecules, and radicals, from where electrons can escape by ionization via interparticle collisions, mainly electron collisions. The plasma served as a large reservoir of free electrons with an energy of approximately 0.95 eV, from where electrons were supplied in the system. First, electrons from the graphite subsurface were transferred to the ionization potential level of EMIM-DCA in water situated at −4.47 eV, which is close to the Fermi level of graphite at −4.19 eV. The electrons in the graphite subsurface were withdrawn, leading to the promotion of peeling graphene layers in graphite. The lack of electrons in few-layer graphene was compensated by an electron from the solution plasma via the electron affinity level of the EMIM dimer in water to few-layer graphene and reached the electronic ground state.

4 Reduction and Oxidation at Plasma-Solid Interface

In the solution plasma, the contact between plasma and solid is the point where the reactions can occur. However, there is still rare evidence of the electron transfer between plasma and solid electrodes in the solution plasma system. Nevertheless, the

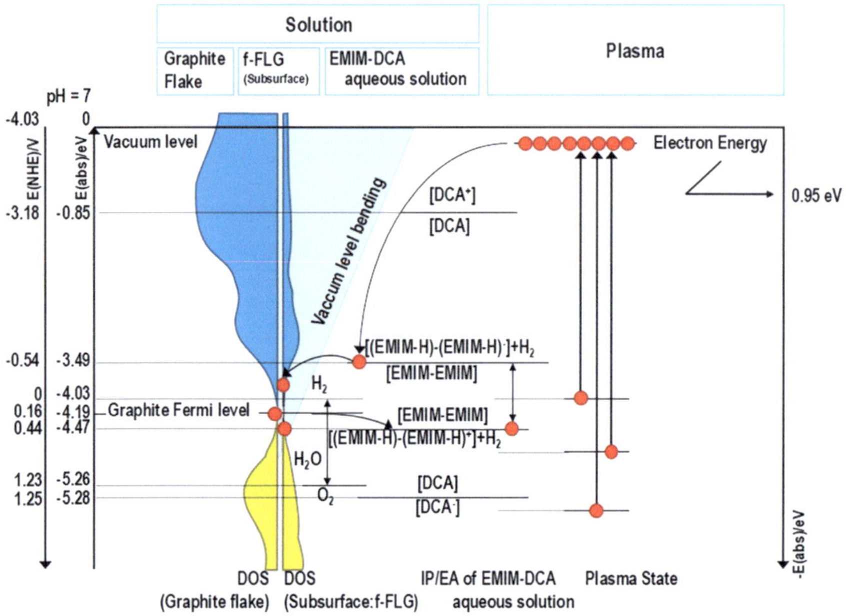

Fig. 12 Illustrating of the few-layer graphene fabrication from graphite flakes by electron exchange induced by the solution plasma. Reprinted with permission from Chae et al. (2017)

investigation of electron transfer between plasma and solid electrodes from the electrical discharge in the gas phase can give some hints. Elahi and co-workers reported an experimental investigation of electron exchange to drive redox reactions in a 10 V potential window at the solid–gas interface by considering gaseous plasma as an electrically conducting medium (Elahi and Caruana 2013). Dynamic electrochemical measurements were conducted in a flame plasma electrolyte in the presence of tungsten oxide salts at 1100 K. Electron transfer and electrochemical reactions at the solid–gas interface were observed using cyclic voltammetry. The effect of working electrode surface area and material, as well as potential scan rate, on the voltammetric redox features, were presented. They described the physical origin of the observed Faradaic current peaks measured in a flame plasma electrolyte and proposed a simple model to describe the reduction–oxidation reaction. There were two aspects of the characteristics that can be clearly extracted. Firstly, there was strong mass transport dependence on the current peaks. Secondly, the reduction processes occurred in both the forward and backward scans at the same potential positions. No reverse oxidation process was observed. The appearance of a peak could be explained by considering an energy level diagram to describe the overlap of electron energy states in the electrode and the gas phase. Electron transfer only occurred when there was a resonance between the Fermi energy, E_F, and the peak reduction potential, Er, of the oxidized states in the gas phase, Fig. 13. Once E_F increased above that of the LUMO, electron

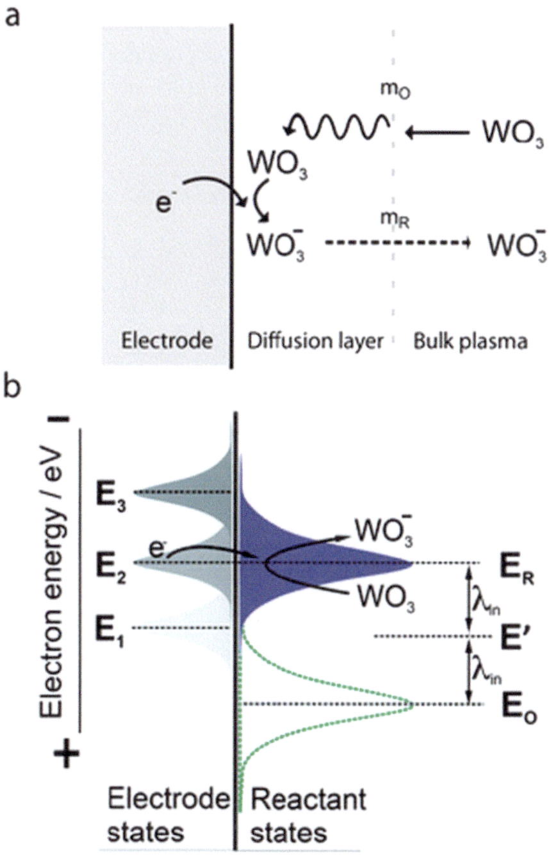

Fig. 13 **a** Summary of mass transport and reaction for WO_3 reduction. The transport of WO_3 was dominated by convection and diffusion, m_O. Whereas WO_3^- was dominated by migration, m_R. **b** Schematic diagram of electron energy levels expressed as available states at three different energy levels and the gas phase states for the reduced, E_R, and oxidized E_{Ol} species. Reprinted with permission from Elahi and Caruana (2013)

transfer increased, as there were no channels to dissipate excess reaction-free energy in the absence of any solvent.

According to the analysis, it could suggest that reduction–oxidation reactions at the solid–gas interface can be the analogous processes at the solid–liquid interface described by conventional electrochemical theory. The departures are mainly due to the mass transport processes that dominate in the gas phase. It can associate migration effects with the total absence of any oxidation processes.

5 Summary

The electron transfer process is the most important process in the solution plasma, which involves the breakdown, propagation, and maintenance of plasma. Moreover, electrons act as a driving force to promote the reactions following the concept of electrochemistry. Even though recently, there have been several studies focusing on plasma electrochemistry in various plasma configurations. The electrochemistry aspect in the solution plasma lacks evidence in both experimental study and calculation. Many phenomena still need to be investigated for the reduction and oxidation reactions by the solution plasma.

References

Arumugam S, Perumal M, Anjana KP, Satyanarayana SVM, Sinha SK (2020) Plasma–metal junction. Phys Plasmas 27(2):023512

Chae S, Bratescu MA, Saito N (2017) Synthesis of few-layer graphene by peeling graphite flakes via electron exchange in solution plasma. J Phys Chem C 121(42):23793–23802

Chen Q, Kaneko T, Matsuda N, Hatakeyama R (2013) Potential structure of discharge plasma inside liquid directly measured by an electrostatic probe. Appl Phys Lett 102(24):244105

Elahi A, Caruana DJ (2013) Plasma electrochemistry: voltammetry in a flame plasma electrolyte. Phys Chem Chem Phys 15(4):1108–1114

Elg DT, Delgado HE, Martin DC, Sankaran RM, Rumbach P, Bartels DM, Go DB (2021) Recent advances in understanding the role of solvated electrons at the plasma–liquid interface of solution-based gas discharges. Spectrochim Acta, Part B 186:106307

Goodman J, Hickling A, Schofield B (1973) The yield of hydrated electrons in glow-discharge electrolysis. J Electroanal Chem Interfacial Electrochem 48(2):319–322

Itikawa Y, Mason N (2005) Cross sections for electron collisions with water molecules. J Phys Chem Ref Data 34(1):1–22

Joshi RP, Thagard SM (2013) Streamer-like electrical discharges in water: Part II. Environmental applications. Plasma Chem Plasma Process 33(1):17–49

Kaneko T, Sasaki S, Takashima K, Kanzaki M (2017) Gas–liquid interfacial plasmas producing reactive species for cell membrane permeabilization. J Clin Biochem Nutr 60(1):3–11

Liu G-H, Wang X-Y, Liu Y-X, Sun J-Y, Wang Y-N (2018) Effects of secondary electron emission on plasma density and electron excitation dynamics in dual-frequency asymmetric capacitively coupled argon plasmas. Plasma Sources Sci Technol 27(6):064004

Mai-Prochnow A, Zhou R, Zhang T, Ostrikov K, Mugunthan S, Rice SA, Cullen PJ (2021) Interactions of plasma-activated water with biofilms: inactivation, dispersal effects and mechanisms of action. npj Biofilms Microbiomes 7(1):11

Morishita T, Ueno T, Panomsuwan G, Hieda J, Yoshida A, Bratescu MA, Saito N (2016) Fastest formation routes of nanocarbons in solution plasma processes. Sci Rep 6:36880

Richmonds C, Witzke M, Bartling B, Lee SW, Wainright J, Liu C-C, Sankaran RM (2011) Electron-transfer reactions at the plasma–liquid interface. J Am Chem Soc 133(44):17582–17585

Saito N, Bratescu MA, Hashimi K (2018) Solution plasma: a new reaction field for nanomaterials synthesis. Jpn J Appl Phys 57(1):0102A4

Shirafuji T, Saito N, Takai O, Tochikubo F, Yasuoka K, Kaneko T, Shirai N (2013) Plasmas in contact with liquid: “Plasma Electrochemistry.” Hyomen Kagaku 34(10):547–552

Witzke M, Rumbach P, Go DB, Sankaran MR (2012) Evidence for the electrolysis of water by atmospheric-pressure plasmas formed at the surface of aqueous solutions. J Phys D Appl Phys 45:44200

Chemical Reactions in Solution Plasma

Nagahiro Saito and Chayanaphat Chokradjaroen

Abstract Solution plasma (SP) is non-equilibrium plasma generated in the liquid phase. Plasma contains electrons of which temperature is as high as 10^4 K. Such high-energy electrons trigger electron impact reactions such as excitation, ionization, and dissociation of molecules and produce active species such as ions, atoms, and radicals. These chemically active species produced in the plasma are supplied to the liquid phase in the SP process (SPP). Chemical reactions at a plasma-liquid interface occur much faster than conventional solution chemistry. Another important feature of the SP is the fact that the plasma is non-equilibrium. It should be noted what the "non-equilibrium" means because this is the most important feature of the SP. In the case of equilibrium plasma, the temperature of the medium may be as high as the electron temperature of 10^3–10^4 K. Such plasma is called thermal plasma, and its use is limited to high-temperature applications such as welding, machining, and waste treatments. In the case of non-equilibrium plasma, by contrast, the temperature of gas and/or liquid is as low as the room temperature even though the electron temperature is as high as 10^4 K. This feature of SPP results in a situation in which active species can participate in the chemistry of solutions without having an excessive thermal effect. This chapter describes various examples of the use of SPPs for materials synthesis.

1 Introduction

This chapter explains the generation of reactive species and chemical reactions generated by the solution plasma. When the voltage is applied, the electrons flow between the metal electrodes submerged in a solution, leading to the collision of particles

N. Saito (✉)
Department of Chemical Systems Engineering, Graduate School of Engineering, Institute for Innovation for Future Society, Nagoya University, Furocho, Chikusa, Nagoya, Japan
e-mail: saito.nagahiro.z7@f.mail.nagoya-u.ac.jp

C. Chokradjaroen
Institute for Aqua Regeneration, Shinshu University, Matsumoto, Nagano, Japan
e-mail: chokradjaroen.c@shinshu-u.ac.jp

N. Saito (ed.), *Solution Plasma Chemistry*, Lecture Notes in Chemistry 115,
https://doi.org/10.1007/978-981-95-0966-9_5

(e.g., charged particles, neutral molecules), sputtering of metal surface, vaporization, and ultraviolet (UV) radiation. Similar to other plasma configurations, the collision of electrons, ions, and a photon with high energy to neutral molecules generally results in excitation, and if the energy is higher than the work function and ionization potential, dissociation and ionization occur. When the energy transfers to the collided molecules, the potential energy of the collided molecules increases and enhances further chemical reactions due to the lowering of overall activation energy. The processes are often called "plasma-activation," which has been of interest in many applications, such as promoting chemical reactions (e.g., ammonia synthesis for energy storage applications (Rouwenhorst et al. 2021), activation of acetaldehyde on the surface of carbon nanotubes for biological application (Chen et al. 2001)), water purification, and other applications.

For dissociation, ionization, and recombination of molecules, the work function, ionization potential, and electron affinity are important parameters. Work function is the energy required to remove an electron from the surface of a metal to the vacuum level outside. In the case of solution plasma, the liquid is assumed to act like a solid semiconductor (as mentioned in Chap. "Electrochemistry in Solution Plasma"). On the other hand, ionization potential is the minimum energy required to remove the outermost electron in an atom in its ground state to infinity, causing the atom to become ionized. Electron affinity refers to the amount of energy released when an electron is attached to a neutral atom or molecule in the gaseous state to form an anion. For instance, of water, the most studied molecule, the interface-specific aqueous-phase work function was 4.73 eV (Thürmer et al. 2021). The ionization potential value of water is ~ 10 eV: according to the study of Delahay and co-workers in 1981, it was 10.06 eV by photoelectron spectroscopy, while more recent photoelectron spectroscopy studies showed a slightly smaller value of 9.9 eV (Delahay and Von Burg 1981; Winter et al. 2004). For electron affinity, a direct measurement of water is still not effective because of the rapid solvation of electrons in water (Gaiduk et al. 2018). The ionization potential value of benzene was reported to be approximately 9.24 eV, while the ions from benzene, such as $C_6H_4^+$ and $C_6H_5^+$, have electron affinity with the averaged value of 13 eV, according to the NIST database.

Moreover, in the solution plasma, phase change is involved, as mentioned in Chap. "What are Solution Plasmas?".The plasma in the gas phase is surrounded by the liquid phase. It can be considered that the solution plasma can provide a heterogeneous catalytic reaction. Within the plasma and at the plasma-liquid interface, the molecules can be excited or go through oxidation/reduction to form a reactive form, which can further react to obtain a product. It is different from the general solid catalyst, which helps to reduce the activation energy.

As mentioned, the plasma and plasma-liquid interface are important reaction sites. Therefore, the reaction routes can be simply divided into two main routes, through the plasma (route 1) and the interface (route 2), as revealed in Fig. 1.

The reaction in the plasma (route 1) should mainly be the reaction through the ionization and dissociation via the collision of electrons and ions with high energy, exceeding an ionization potential, to molecules. The reactions at the plasma-liquid interface (route 2) can also occur by the electron transfer or collision of electrons

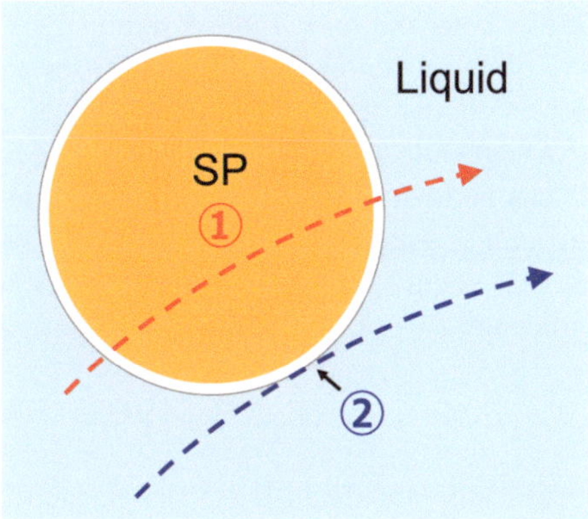

Fig. 1 Reaction routes in solution plasma through plasma zone (route 1) and plasma-liquid interface (route 2)

and ions, which sometimes contain lower energy than that in the plasma. Morishita and co-workers confirmed the presence of the two routes (Morishita et al. 2016). The reaction route through plasma was found to consist of the vaporized and decomposed species, and the plasma temperature was as high (~4000 K) as that of the pyrolysis process. The reaction route at the interface between the plasma and solution, where the ion temperature is as low as room temperature, but the electron temperature is still sufficient to proceed with the reaction. Moreover, the solution plasma was proposed to have a reaction window with an average electron energy of approximately 7 eV (Chae et al. 2017). Consequently, it was suggested that electrons could be exchanged at the interface and carry on the electrochemical reaction (Chae et al. 2017), as explained in Chap. "Electrochemistry in Solution Plasma". For the electrochemical reaction, the reduction–oxidation potential must be considered to find the tendency of a molecule to acquire electrons from or lose electrons to an electrode to succeed in the reduction–oxidation reaction, respectively.

2 Chemical Reactions in Aqueous Solution by Solution Plasma

Solution plasma has been proposed to induce a variety of reactive species in an aqueous solution, which has water molecules as a dominant specie and other molecules dissolved in water (H_2O), such as oxygen (O_2). These reactive species include both short-lived reactive species (e.g., hydroxyl radical ($^{\bullet}OH$) and hydrogen radical ($^{\bullet}H$)) and long-lived reactive species (e.g., hydrogen peroxide (H_2O_2)), which can facilitate several reactions, such as degradation of organic compounds, without the addition of chemical agents. The formation of each reactive species is different, depending on reaction zones and the type of reactions. When the H_2O molecule enters the plasma area (route 1), it is vaporized due to the high plasma temperature. The vaporized H_2O molecule can be excited, dissociated, and ionized via the collision of electrons and ions. Table 1 summarizes the threshold energies through previous works (Itikawa and Mason 2005; Joshi and Thagard 2013) as a reference for the excitation, dissociation, and ionization of water molecules.

Meanwhile, the reaction of an interface between the plasma and liquid water (route 2) can be altered depending on the species and their work function. Therefore, in the case of water molecules with a work function of 4.73 eV, the electron with high enough energy at the interface is sufficient to remove the electron from the water molecule:

$$H_2O \rightarrow H_2O^{\bullet +} + e^-$$

The solution chemistry of the plasma-liquid interface for the solution plasma is still unclear. However, the plasma-liquid interface has been intensively interested in and investigated in solution-based gas discharge, which can be used to understand the plasma-liquid interface of solution plasma. Rumbach and co-workers developed a total internal reflection absorption spectroscopy (TIRAS) technique to observe electrons at the plasma-liquid interface. Briefly, TIRAS applies an incident laser beam penetrating through the liquid at a shallow angle and below the critical angle, and the laser light is entirely reflected at the liquid–air interface. Its intensity is measured, and the difference from the no plasma condition is assumed to be light absorbed by an electron at the interface and shallow liquid bulk (Rumbach et al. 2015). The result exhibited that the solvated electron, $e^-_{(aq)}$, and hydroxyl anion, $OH^-_{(aq)}$, are primary species in the interfacial regions. Furthermore, it was also reported that

Table 1 Reactions via the collision of electrons and neutral water molecules

Type of reaction	Reaction	Threshold (eV)	Solution plasma
Momentum transfer	$H_2O + e^- \rightarrow H_2O + e^-$	0	
Rotational excitation	$H_2O + e^- \rightarrow H_2O^* + e^-$	0.0009–0.198	
Vibrational excitation	$H_2O + e^- \rightarrow H_2O^* + e^-$	0.198–0.453	
Dissociation	$H_2O + e^- \rightarrow {}^{\bullet}OH + {}^{\bullet}H + e^-$	7.0–19.0	(×)
Dissociation	$H_2O + e^- \rightarrow H_2 + O^* + e^-$	17.0–19.0	×
Dissociation	$H_2O + e^- \rightarrow H_2 + {}^{\bullet}O + e^-$	23.5	×
Dissociation	$H_2O + e^- \rightarrow OH^* + H$	9.0	×
Ionization	$H_2O + e^- \rightarrow 2e^- + H_2O^+$	13.0	×
Ionization	$H_2O + e^- \rightarrow 2e^- + {}^{\bullet}H + HO^+$	15.0	×
Ionization	$H_2O + e^- \rightarrow 2e^- + H_2 + O^+$	22.5	×
Ionization	$H_2O + e^- \rightarrow 2e^- + H_2^+ + O$	29.0	×
Ionization	$H_2O + e^- \rightarrow 2e^- + H^+ + {}^{\bullet}OH$	19.0	×
Attachment	$H_2O + e^- \rightarrow OH^- + {}^{\bullet}H$	4.30	
Attachment	$H_2O + e^- \rightarrow H^- + {}^{\bullet}OH$	4.36	
Attachment	$H_2O + e^- \rightarrow H_2 + O^-$	4.43	

the presence of $OH^-_{(aq)}$ in a large number could be produced from the highly reactive hydroxyl radical, $^\bullet OH$, generated near the interface through:

$$e^-_{(aq)} + {}^\bullet OH \rightarrow OH^-_{(aq)}$$

Accordingly, the electrochemical approaches can also be used to explain the reactions at the plasma-liquid interfaces in the aqueous solution (Buxton et al. 1988). In applications, it might be sufficient that electrons solvated in water molecules can recombine at the interface through an interface neutralization mechanism, with, for example, ions reaching the interface from the plasma (Bruggeman et al. 2016),

$$e^-_{(aq)} + A^+ \rightarrow A + h\nu.$$

This reaction can be helpful for the reduction of metal nanoparticles, which will be discussed in the next section. Besides, there are several possible reactions by the solvated electrons, as shown in Table 2. The actual reactions were proposed to depend upon the number of solvated electrons and other ions, temperature, pH, and dissolved gases (Norberg et al. 2014). For example, a modification of biopolymers, e.g., cellulose, chitosan, and alginate, is one of the reactions that have been generally conducted by solution plasma in aqueous solutions (Chokradjaroen et al. 2020; Davoodbasha et al. 2018; Prasertsung et al. 2012; Tantiplapol et al. 2015). These biopolymers are polysaccharides, long-chain polymeric carbohydrates composed of monosaccharide units bound together by glycosidic linkages. Although there is still unclear evidence, it was proposed that the $^\bullet OH$ plays an important role in the degradation through the CH-activation near the glycosidic linkages. The reaction can lead to the chain scission without altering the backbone structure of chitosan. The reaction should carry on at the plasma-liquid interface.

Table 2 The products obtained from the linear structures with symbolic numbers corresponding to the GC/MS spectrograms

C numbers	Strucutres with Symbolic Numbers for Liniar Molecules			
C4	L1)	1-Butene		
C5	L2)	1-Pentene		
C6	L3)	1-Hexene		
C7	L4)	1-Heptene	L10)	Heptane
C8	L5)	1-Octene	L11)	Heptane
C9	L6)	1-Nonene	L12)	Nonane
C10	L7)	1-Decene	L13)	Decane
C11	L8)	1-Undecene	L14)	Undecane
C12	L9)	1-Dodecene	L15)	Dodecane

Reprinted with permission from Morishita et al. (2016)

3 Chemical Reactions in Organic Solution by Solution Plasma

The solution plasma reaction in organic solutions (organic solvents) has been studied intensively, yielding carbon materials potentially useful for energy applications such as lithium-ion batteries. The formation routes of carbon nanoparticles were examined by Morishita and co-workers (Morishita et al. 2016). These were investigated using hexane, hexadecane, cyclohexane, and benzene as carbon precursors. It was found that there were two routes, depending on the chemical structure of carbon precursors. Linear molecules decomposed into shorter olefins, such as C_2, in the plasma area and then went through pyrolysis synthesis due to the high temperature of plasma (Fig. 2). For the saturated ring molecules, C–H dissociation proceeded at the plasma-liquid interface, followed by conversion into unsaturated ring molecules via hydrogenation. Meanwhile, unsaturated ring molecules lost electrons at the plasma-liquid interface and transformed into cation radicals, such as benzene radical cation, which were directly polymerized into two- and three-ring molecules, which are intermediates (Fig. 3). The information of formed radicals and intermediates formed in the solution plasma-induced formation of carbon nanoparticles is shown in Table 2 and Table 3, respectively. According to the different precursors and routes, the carbon nanoparticles from linear molecules were more crystalline than those from ring molecules. Moreover, unsaturated ring molecules, i.e., benzene, could yield the highest amount of carbon nanoparticles Fig. 4. The evidence could indicate that the rate of reaction through the electron exchange at the plasma-liquid interface is relatively high, owing to the high number of electrons, presented in the plasma-liquid interface, and the larger reaction area, compared to plasma area.

The formation of carbon nanoparticles at the plasma-liquid interface, in which the reaction condition is maintained at atmospheric pressure and room temperature, could be suggested to be suitable and has been used for the in-situ synthesis of

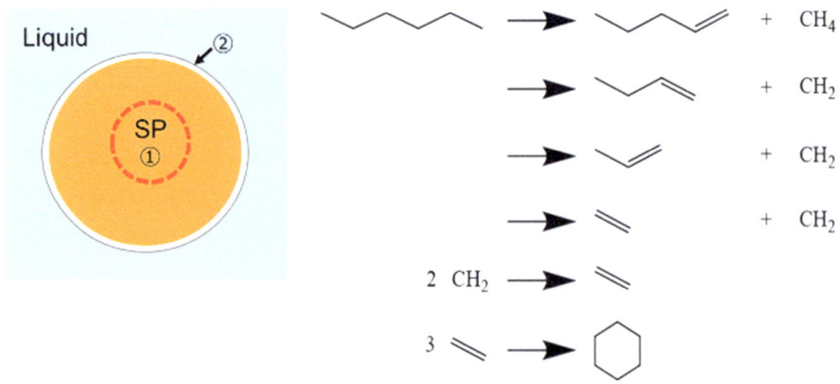

Fig. 2 The primary reactions of linear molecules proceed via thermal reactions in the plasma area (Morishita et al. 2016)

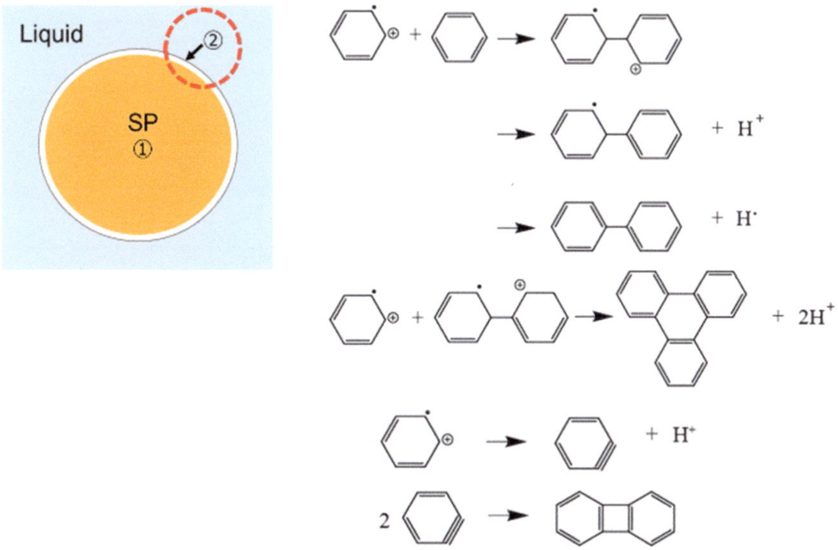

Fig. 3 The reactions of unsaturated ring molecules proceed via polymerization reactions at the plasma-liquid interface area (Morishita et al. 2016)

heteroatom-doped carbon nanoparticles. For example, a novel cationic nitrogen-doped graphene by the solution plasma was proposed by Saito and co-workers (Chae et al. 2019). A mixture of 10% 1-ethyl-3-methylimidazolium dicyanamide (EMIM DCA, $C_8H_{11}N_5$) and 90% dimethylformamide (DMF, C_3H_7NO) was used as a precursor. The presumable formation mechanism of cationic nitrogen-doped graphene is schematically illustrated in Fig. 5. Each EMIM cation is first bound to another one through the CH activation at the plasma-liquid interface. The combined two EMIM cations are subsequently oxidized, which makes all of the N contained in the EMIM cations transform into cationic N. Due to the instability of π-conjugated bonds in the five-membered (5) ring, they transform from the 5 + 5 ring to the 6 + 6 ring through the inverse Stone–Wales (ISW) reaction. The ISW reaction of EMIM cations contributes mainly to the formation of CNG. In addition, the EMIM DCA and DMF can be dissociated into free C–C, C–N, and C–N–C through the plasma area, and it is assembled with sp2 hybridized carbon containing N for graphene sheet assembly. The doping percent was 13.4 atom %, which is relatively high, compared to that of other methods (Lemes et al. 2019; Tung et al. 2018).

Table 3 The products obtained from the ring structures with symbolic numbers corresponding to the GC/MS spectrograms

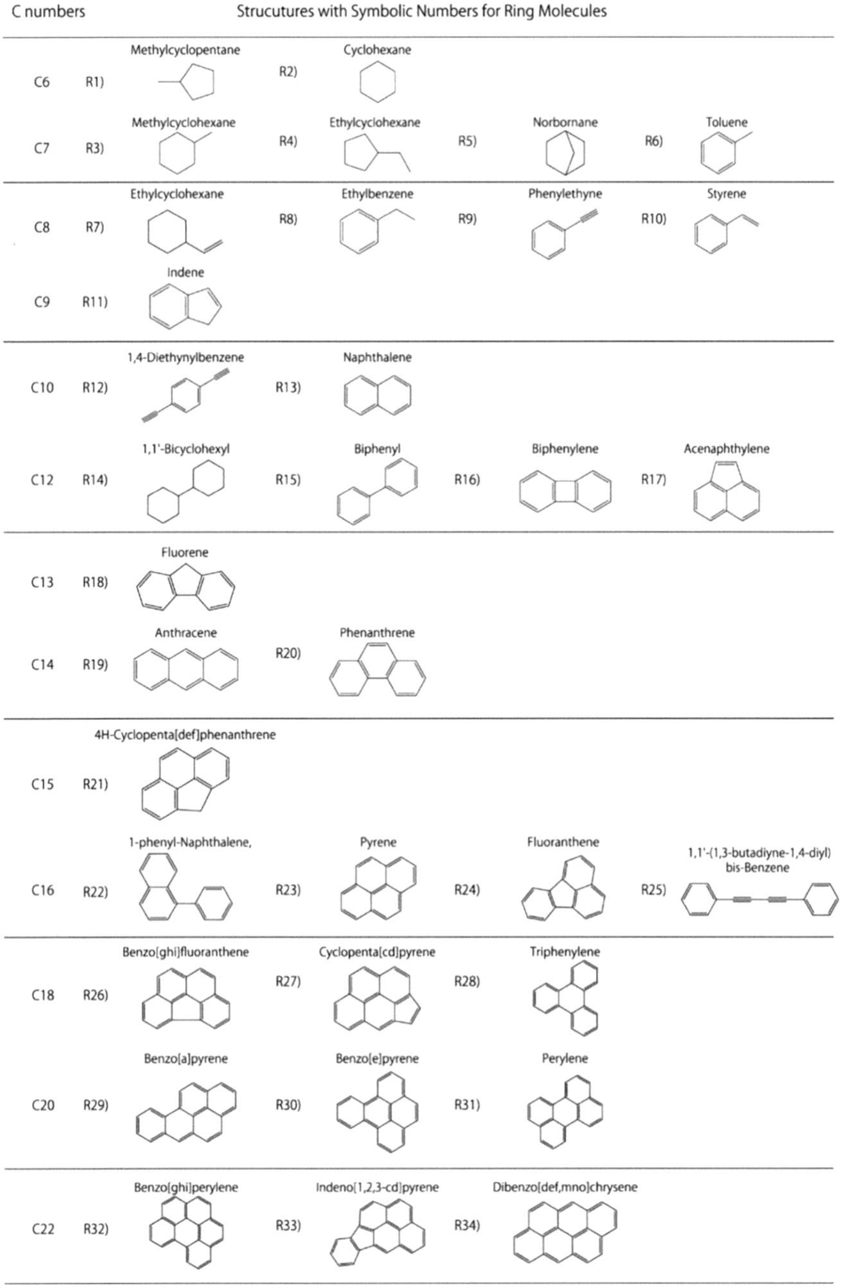

Reprinted with permission from Morishita et al. (2016)

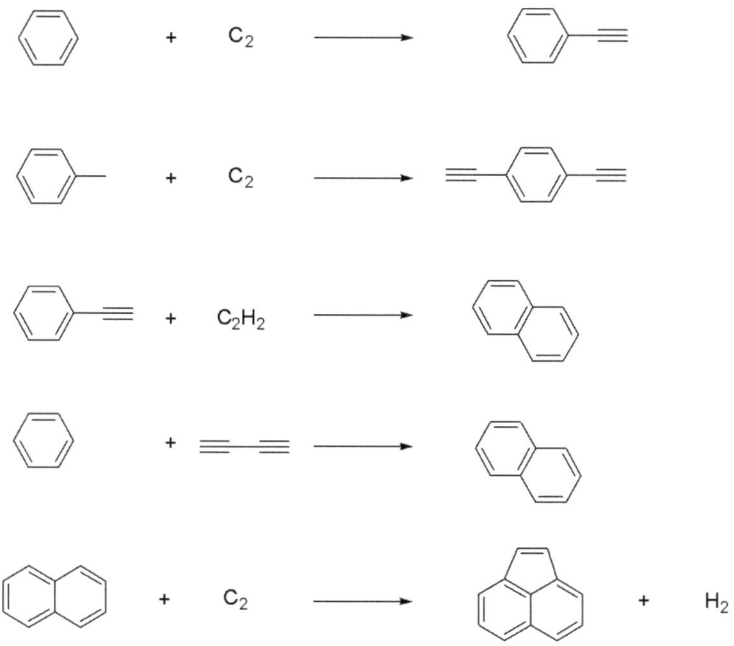

Fig. 4 The side reactions of unsaturated ring molecules and decomposed molecules proceed at the plasma-liquid interface area (Morishita et al. 2016)

4 Reduction Reaction of Metal Ions via Solution Plasma

Solution plasma is a useful and straightforward method for the synthesis of metal nanoparticles because non-equilibrium plasma can rapidly produce several reactive species. Moreover, the plasma-liquid interface can provide the reaction field at atmospheric pressure and low temperature, almost room temperature. Saito and co-workers applied the solution plasma to synthesize gold nanoparticles in a water-based system. It was found that $^{\bullet}H$ species were responsible for the gold nanoparticles formation through the reduction of $[AuCl_4]^-$ ion (Saito et al. 2009). Compared with ordinary chemical reduction, the $^{\bullet}H$ plays the role of a reducing agent and provides electrons to a metal ion. As the discharge time is prolonged, $^{\bullet}H$ is continuously generated; the solution plasma achieves the high-speed synthesis of metal nanoparticles. The reaction mechanism of the gold nanoparticle formation by the solution plasma is clearly explained: In the initial stage of the discharge, $[AuCl_4]^-$ ion is reduced, and dendrite-shaped nanoparticles with a size less than 1 nm, due to the nucleation reaction, are concurrently synthesized to form spherical nanoparticles. Then, the pH of the solution gradually decreases with increasing the discharge time, leading to the dissolution of gold nanoparticles at low pH. The rate of reduction of $[AuCl_4]^-$ ion declines.

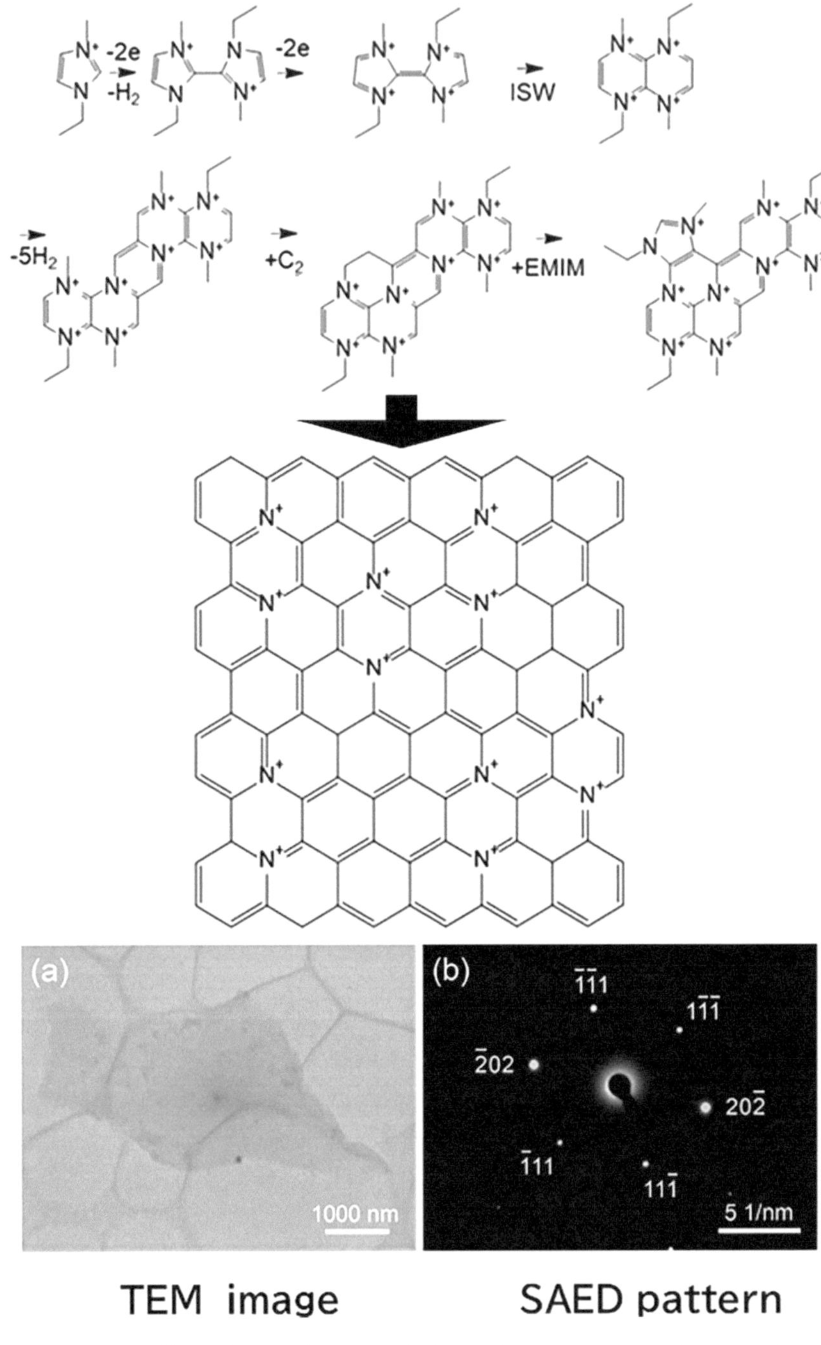

Fig. 5 Schematic illustration showing the synthesis of cationic nitrogen-doped graphene from the mixture of EMIM DCA and DMF by solution plasma and its TEM image and SAED pattern. Reprinted with permission from Chae et al. (2019)

Gold nanoparticles with anisotropic shapes form in the solution through the etching process. Later, Bratescu and co-workers demonstrated that the solution pH could be adjusted to control the size of the synthesized gold nanoparticles by the solution plasma under an aqueous environment (Bratescu et al. 2011). The gold nanoparticles exhibit sizes from 1 to 2 to 10 nm when the solution pH was adjusted in the range from 12 to 3, respectively. According to the plasma and product characterization, the reaction mechanism was proposed as follows:

$$H_2O \rightarrow {}^{\bullet}H + {}^{\bullet}OH$$

$$2H_2O \rightarrow O_2 + 2H_2$$

$$O_2 \rightarrow 2O^{\bullet}$$

$$H_2 \rightarrow 2H^{\bullet}$$

$$H + e^- \rightarrow H^* + e^-$$

$$O + e^- \rightarrow O^* + e^-$$

where $H^{\bullet}$ and $O^{\bullet}$ mean radical, H^* and O^* mean excited atom, respectively, and e is the electron from plasma. $H^{\bullet}$ mainly exists in the gas phase, consistent with Fig. 2; therefore, the reduction of Au^{3+} to atomic Au^0 should proceed at the plasma-liquid interface. The reactions could occur depending on the solution's pH, as shown below:

$$AuCl_4{}^- + OH^- \leftrightarrow Au(OH)_4{}^- + Cl^-$$

$$AuCl_4{}^- + 3H^{\bullet} \leftrightarrow HCl + Au^0 + Cl^- \ (pH\ 3)$$

$$Au(OH)_4{}^- + 3H^{\bullet} \leftrightarrow 3H_2O + Au^0 + Cl^- (pH\ 12).$$

Furthermore, Sudare and co-workers demonstrated using an organic–water mixture as a liquid medium for the enhanced synthesis of gold nanoparticles by the solution plasma (Sudare et al., 2015). The presence of organic solvents, such as ethanol, could increase the reaction rate by ~35.2 times faster than that in pure water. The dependence of these changes on ethanol concentration, the period when the concentration of gold ions (A) had decreased to half the value of the initial concentration (A_0), $t_{1/2}$, was plotted against mole fractions of ethanol in Fig. 6. This result showed that, at mole fractions of ethanol equaled to 0.089, the fastest reaction rate could be observed, and the partial molar volumes of ethanol and water were at their minimum and maximum, respectively. Consequently, it could be indicated that H$^{\bullet}$ generated in the plasma immediately reacts with alcohol molecules at the plasma-liquid interface and/or in bulk solution to form alcohol radicals, which is more stable in the viewpoint of such reactivity. Besides, alcohol itself is known to behave as a reducing agent and has been used for nanoparticle synthesis. Therefore, the presence of alcohol could lead to a synergetic influence on the synthesis of metal nanoparticles via the solution plasma, and the reaction could be proposed as follows:

$$CH_3CH_2OH + {}^{\bullet}H \rightarrow {}^{\bullet}CH(CH_3)OH + H_2$$

$$CH_3CH_2OH + {}^{\bullet}OH \rightarrow {}^{\bullet}CH(CH_3)OH + H_2O$$

$$AuCl_4{}^{-} + 3CH(CH_3)OH \rightarrow Au^0 + 3CH_3CHO + 3HCl + Cl^{-}$$

Moreover, since the solution plasma system is discharged inside the liquid phase, the generation of hydrated electrons, $e^{-}{}_{(aq)}$, should be possible, similar to the plasma-liquid interface in other solution-based gas discharge (Elg et al. 2021). Sankaran and co-workers showed evidence that $e^{-}{}_{(aq)}$ in a negatively biased plasma could reduce metal cations to zero-valent metal, e.g., silver (Ag) (Ghosh et al. 2017). The reduction of metal cations, e.g., silver cations (Ag^{+}), by using the hydrated electrons was proposed to analog an electroplating process, as shown below:

$$e^{-}_{(aq)} + Ag^{+} \rightarrow Ag^{0}.$$

However, unlike electroplating, there is no solid electrode because the metal atoms form colloidal nanoparticles. According to this evidence, the solution plasma possibly provides two routes for the reduction reaction of metal ions, as revealed in Fig. 7.

5 Sputtering of Metal Surface via Solution Plasma

In the solution plasma, the collision of electrons and ions on the metal surface of the electrodes is a very important process for generating secondary electrons to sustain the non-equilibrium plasma. Nevertheless, when collisions of ions with enough energy on the metal surfaces occur, they can lead to the ejection of metal atoms from the surface. It becomes a benefit for the synthesis of metal nanoparticles without the addition of chemicals, such as precursors and reducing agents. The process is sometimes called solution plasma sputtering (SPS). Sputtering of the metal surface occurs by the collisions of ions with high energy on solid surfaces, such as carbon and metal surfaces, leading to the ejection of atoms (i.e., ion bombardment). The ejected atoms subsequently formed clusters and nucleation and later nanoparticles, as shown in Fig. 8. For example, Hu and co-workers could produce well-crystallized platinum-gold alloy nanoparticles via solution plasma sputtering using platinum and gold electrodes (Hu et al. 2013). The graphite electrodes were also proposed in the

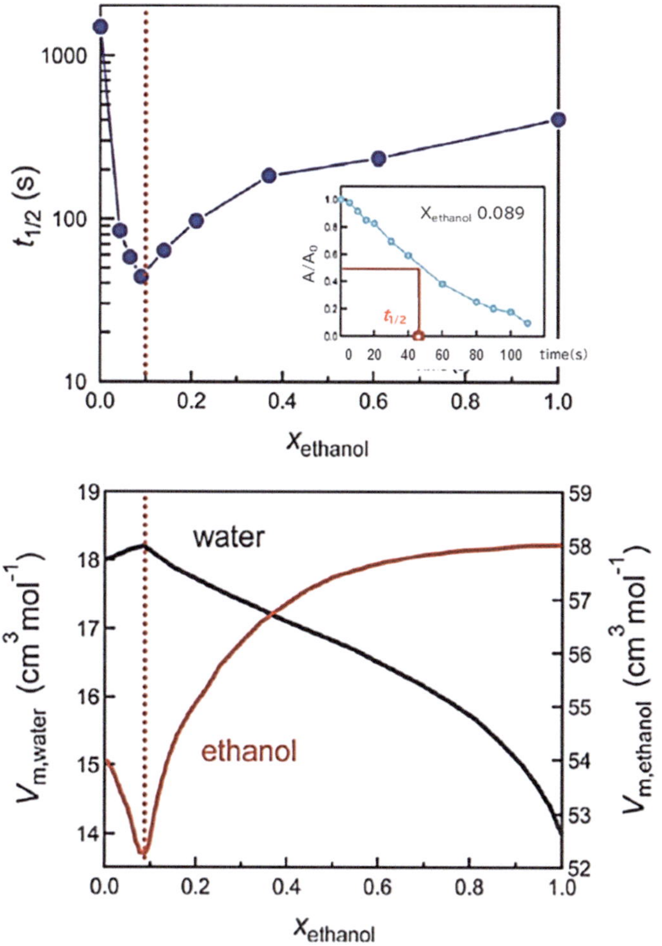

Fig. 6 The correlation between the $t_{1/2}$ value and the partial molar volumes of water and ethanol, respectively, in ethanol–water mixtures as a function of the mole fractions of ethanol, $\chi_{ethanol}$. Reprinted with permission from Sudare et al. (2015)

study by Saito and co-workers that they can be used in the solution plasma sputtering to produce graphene (Chae et al. 2017).

SPS directly relates to the solid surface and ion bombardment. The types of solid surfaces and ions are required to be considered. In general, in the sputtering process, one of the requirements for solid materials is favorable work functions. Work function can be defined as the minimum energy required to extract an electron from the surface of a solid where an atom can be simultaneously rejected. This minimum energy refers to the energy that exists in the ions and participates in the collision on the solid surface. It also relates to the potential and kinetic energy of charged particles (e.g., mass). Therefore, the ability of the nanoparticle formation

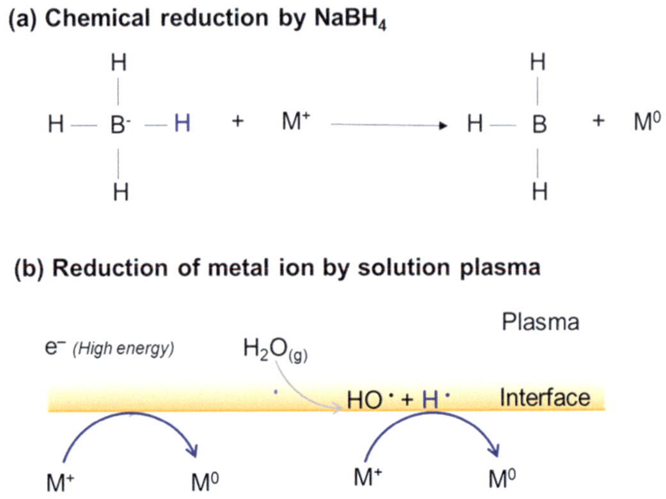

Fig. 7 **a** Chemical reduction reaction of metal ions (M^+) to metal nanoparticles (M^0) by using sodium borohydride ($NaBH_4$) as reducing agents. **b** Reduction reaction of metal ions (M^+) to metal nanoparticles (M^0) by hydrogen radical ($^{\bullet}H$) and electron (e^-), generated by solution plasma

Fig. 8 Illustration of solution plasma sputtering on the metal surface of electrodes

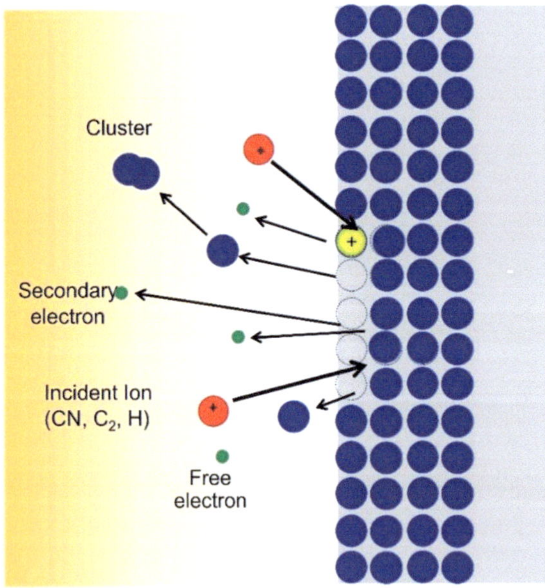

Table 4 Example of electrode materials for solution plasma with their work function values (Chae et al. 2017; "CRC Handbook of Chemistry and Physics," 1977)

Chemical element	Work function (eV)
Tungsten	4.30–5.22
Silver	4.26–4.74
Iron	4.67–4.81
Gold	5.10–5.47
Graphite	4.19
Titanium	4.33
Zinc	3.63–4.9
Platinum	5.12–5.93
Nickel	5.04–5.35
Palladium	5.22–5.60

(e.g., size, size distribution, and production rate) in the solution plasma sputtering differs depending on the work functions of solid electrodes and the type of solutions that determine the reactive species. Table 4 summarizes the work function of solid materials used as electrodes for the solution plasma sputtering.

6 Summary

Solution plasma is a non-thermal equilibrium plasma that can provide rapid and continuous reactions. The reaction fields in the solution plasma are divided into two areas, plasma and plasma-liquid interface, which can promote different reactions. In the plasma area, the electron and ion collisions drive the decomposition of molecules, and ion bombardments cause the metal sputtering to produce nanomaterials. Therefore, the ionization potential and work function of molecules are important parameters. Meanwhile, the plasma-liquid interface possibly acts like another solid electrode that can carry on the reaction under the electrochemical approach. However, several phenomena still need to be investigated to provide accurate and clear explanations of the reactions induced by solution plasma.

References

Bratescu MA, Cho S-P, Takai O, Saito N (2011) Size-controlled gold nanoparticles synthesized in solution plasma. J Phys Chem C 115(50):24569–24576

Bruggeman PJ, Kushner MJ, Locke BR, Gardeniers JGE, Graham WG, Graves DB, Hofman-Caris RCHM, Maric D, Reid JP, Ceriani E, Fernandez Rivas D, Foster JE, Garrick SC, Gorbanev Y, Hamaguchi S, Iza F, Jablonowski H, Klimova E, Kolb J, Krčma F, Lukes P, Machala Z, Marinov I, Mariotti D, Mededovic Thagard S, Minakata D, Neyts EC, Pawłat J, Petrović ZL, Pflieger R, Reuter S, Schram DC, Schröter S, Shiraiwa M, Tarabová B, Tsai PA, Verlet JRR, von Woedtke

T, Wilson KR, Yasui K, Zvereva G (2016) Plasma–liquid interactions: a review and roadmap. Plasma Sources Sci Technol 25(5):053002
Buxton GV, Greenstock CL, Helman WP, Ross AB (1988) Critical review of rate constants for reactions of hydrated electrons, hydrogen atoms and hydroxyl radicals ($\cdot OH/\cdot O^-$) in aqueous solution. J Phys Chem Ref Data 17(2):513–886
Chae S, Bratescu MA, Saito N (2017) Synthesis of few-layer graphene by peeling graphite flakes via electron exchange in solution plasma. J Phys Chem C 121(42):23793–23802
Chae S, Panomsuwan G, Bratescu MA, Teshima K, Saito N (2019) P-type doping of graphene with cationic nitrogen. ACS Appl Nano Mater 2(3):1350–1355
Chen Q, Dai L, Gao M, Huang S, Mau A (2001) Plasma activation of carbon nanotubes for chemical modification. J Phys Chem B 105(3):618–622
Chokradjaroen C, Rujiravanit R, Theeramunkong S, Saito N (2020) Effect of electrical discharge plasma on cytotoxicity against cancer cells of N, O-carboxymethyl chitosan-stabilized gold nanoparticles. Carbohydr Polym 237:116162
Davoodbasha M, Lee S-Y, Kim J-W (2018) Solution plasma mediated formation of low molecular weight chitosan and its application as a biomaterial. Int J Biol Macromol 118:1511–1517
Delahay P, Von Burg K (1981) Photoelectron emission spectroscopy of liquid water. Chem Phys Lett 83(2):250–254
Elg DT, Delgado HE, Martin DC, Sankaran RM, Rumbach P, Bartels DM, Go DB (2021) Recent advances in understanding the role of solvated electrons at the plasma–liquid interface of solution-based gas discharges. Spectrochim Acta, Part B 186:106307
Gaiduk AP, Pham TA, Govoni M, Paesani F, Galli G (2018) Electron affinity of liquid water. Nat Commun 9(1):247
Ghosh S, Hawtof R, Rumbach P, Go DB, Akolkar R, Sankaran RM (2017) Quantitative study of electrochemical reduction of Ag^+ to Ag nanoparticles in aqueous solutions by a plasma cathode. J Electrochem Soc 164(13):D818–D824
Hu X, Shen X, Takai O, Saito N (2013) Facile fabrication of PtAu alloy clusters using solution plasma sputtering and their electrocatalytic activity. J Alloy Compd 552:351–355
Itikawa Y, Mason N (2005) Cross sections for electron collisions with water molecules. J Phys Chem Ref Data 34(1):1–22
Joshi RP, Thagard SM (2013) Streamer-like electrical discharges in water: Part II. Environmental applications. Plasma Chem Plasma Process 33(1):17–49
Lemes G, Sebastián D, Pastor E, Lázaro MJ (2019) N-doped graphene catalysts with high nitrogen concentration for the oxygen reduction reaction. J Power Sources 438:227036
Morishita T, Ueno T, Panomsuwan G, Hieda J, Yoshida A, Bratescu MA, Saito N (2016) Fastest formation routes of nanocarbons in solution plasma processes. Sci Rep 6:36880
Norberg SA, Tian W, Johnsen E, Kushner MJ (2014) Atmospheric pressure plasma jets interacting with liquid covered tissue: touching and not-touching the liquid. J Phys D Appl Phys 47(47):475203
Prasertsung I, Damrongsakkul S, Terashima C, Saito N, Takai O (2012) Preparation of low molecular weight chitosan using solution plasma system. Carbohydr Polym 87(4):2745–2749
Rouwenhorst KHR, Burbach HGB, Vogel DW, Núñez Paulí J, Geerdink B, Lefferts L (2021) Plasma-catalytic ammonia synthesis beyond thermal equilibrium on Ru-based catalysts in non-thermal plasma. Catal Sci Technol 11(8):2834–2843
Rumbach P, Bartels DM, Sankaran RM, Go DB (2015) The solvation of electrons by an atmospheric-pressure plasma. Nat Commun 6:7248
Saito N, Hieda J, Takai O (2009) Synthesis process of gold nanoparticles in solution plasma. Thin Solid Films 518(3):912–917
Sudare T, Ueno T, Watthanaphanit A, Saito N (2015) Accelerated nanoparticles synthesis in alcohol–water-mixture-based solution plasma. Phys Chem Chem Phys 17(45):30255–30259
Tantiplapol T, Singsawat Y, Narongsil N, Damrongsakkul S, Saito N, Prasertsung I (2015) Influences of solution plasma conditions on degradation rate and properties of chitosan. Innov Food Sci Emerg Technol 32:116–120

Thürmer S, Malerz S, Trinter F, Hergenhahn U, Lee C, Neumark DM, Meijer G, Winter B, Wilkinson I (2021) Accurate vertical ionization energy and work function determinations of liquid water and aqueous solutions. Chem Sci 12(31):10558–10582

Tung TT, Alotaibi F, Nine MJ, Silva R, Tran DNH, Janowska I, Losic D (2018) Engineering of highly conductive and ultra-thin nitrogen-doped graphene films by combined methods of microwave irradiation, ultrasonic spraying and thermal annealing. Chem Eng J 338:764–773

Winter B, Weber R, Widdra W, Dittmar M, Faubel M, Hertel IV (2004) Full valence band photoemission from liquid water using EUV synchrotron radiation. J Phys Chem A 108(14):2625–2632

Materials Synthesis Using the Solution Plasma Reaction

Nagahiro Saito and Chayanaphat Chokradjaroen

Abstract Solution plasma processes (SPPs) have possibility to be applied to various fascinating applications, especially to the synthesis of nanomaterials. These studies have been conducted with small laboratory-level experimental equipment. In such equipment, plasma is generated in the gap (approx. 1 mm) between a pair of electrodes (diam. approx. 1 mm). The size of plasma is hence mm scale. This size of plasma is quite smaller than the volume of the liquid to be treated even if we liquid is as small as a few 10 mL. In the case of industrial application, the volume of liquid to be treated can be few 10 L a day. Thus, if you apply SPPs to an industrial application, you must develop a suitable reactor for processing large amounts of liquid. This chapter describes several efforts toward industrial SPPs.

1 Introduction

Solution plasma (SP) is non-equilibrium plasma generated in the liquid phase. Plasma contains electrons whose temperature is as high as 10^4 K. Such high-energy electrons trigger electron impact reactions such as excitation, ionization, and dissociation of molecules and produce active species such as ions, atoms, and radicals (Lieberman and Lichtenberg 2005; Fridman 2008). These chemically active species produced in the plasma are supplied to the liquid phase in the SP process (SPP). Chemical reactions at a plasma-liquid interface occur much faster than in conventional solution chemistry. Another important feature of the SP is the fact that the plasma is non-equilibrium. What "non-equilibrium" means should be noted because this is the most important feature of the SP. In the case of equilibrium plasma, the temperature of the medium may be as high as the electron temperature of 10^3–10^4 K. Such plasma is

N. Saito (✉)
Institute for Innovation for Future Society; Department of Chemical Systems Engineering, Graduate School of Engineering, Nagoya University, Nagoya, Japan
e-mail: saito.nagahiro.z7@f.mail.nagoya-u.ac.jp

C. Chokradjaroen
Institute for Aqua Regeneration, Shinshu University, Nagano, Japan
e-mail: chokradjaroen.c@shinshu-u.ac.jp

N. Saito (ed.), *Solution Plasma Chemistry*, Lecture Notes in Chemistry 115,
https://doi.org/10.1007/978-981-95-0966-9_6

Table 1 Applications of SPPs

NP	Soution	Plasma	References	Section
Au	$HAuCl_4 \cdot 4H_2O$, $HAuCl_4 + SDS$, $HAuCl_4 \cdot [BMIM] [BF_4]$	Pulse SP, GDE	Hieda et al. (2008a, b), Kim et al. (2008)	6.3.1 6.3.2
Ag	$AgNo_3$ solution, $AgNo_3 + PVP$	Pulse SP, MWSP	Pootawang et al. (2011a, b) Ashkarran (2010)	6.3.3 6.3.7
Pd	$PdCl_2 + [BMIM] [BF_4] +$ HCl	LPGD	Xie and Liu (2008)	6.5
Pt	H_2PtCl_6 solution	GDE	Lal et al. (2008) Koo et al. (2005)	6.3.4
Cu	$CuCl_2$ + ascorbic acid, $CuSO_4 + H_2SO_4$	Pulse SP, GDE	Pootawang et al. (2013a, b) Lal et al. (2008)	6.4
NI	$NICI_2$, LICI–KCI–CsCI	GDE	Tokushige et al. (2009a, b)	6.4
Fe	$FeCI_2$, LICI–KCI–CsCI	GDE	Tokushige et al. (2010a, b)	6.4

called thermal plasma, and its use is limited to high-temperature applications such as welding (Bohnart 2018), machining (Jameson 2001; Kibria et al. 2019), and waste treatments (Heberlein and Murphy 2008; Safa and Soucy 2014). In the case of non-equilibrium plasma, by contrast, the temperature of a gas and/or liquid is as low as the room temperature even though the electron temperature is as high as 10^4 K. This feature of SPP results in a situation in which active species can participate in the chemistry of solutions without having an excessive thermal effect. Possible applications are listed in Table 1. This chapter describes various examples of the use of SPPs for materials synthesis. Table 1 guides the topics covered in this chapter.

where GDE means glow discharge electrolysis, MWSP means microwave solution plasma, and LPGD is low-pressure glow discharge.

2 Nanomaterials Synthesis

Currently, the most frequently reported application of SPPs is the synthesis of nanomaterials. Nanomaterials are defined as clusters of atoms or molecules measured in the nanoscale size range, which is between 1 and 100 nm, according to the International Organization for Standardization (ISO) (2015). Various nanomaterials have been synthesized by SPPs to date. Table 2 shows several examples of nanoparticles synthesized by SPPs. As will be introduced later in this chapter, many other materials have been synthesized by SPPs. Why are so many nanomaterials synthesized using SPPs? The answer to this question is that SPPs can be conducted under the following fascinating conditions:

Table 2 Examples of nanoparticles synthesized by SPPs

Process	Materials	Applications	Section
Nanoparticle synthesis	Metal nanoclusters, Nanoparticles supported on mesoporous silica	Fuel cell catalysts, Environmental catalysts, Battery materials Electrode pastes, Biosensors	6.3 6.4 6.5
Surface modification	Carbon nanotube link, Functional graphene, Functional polymer	Polymer composites, Composite coatings, Electrode materials, Transparent electrodes, Food chemicals	6.6 6.7 6.8
Carbon synthesis	Mesoporous carbon, Graphene flakes, Hetero carbon	Fuel cell catalysts, Battery catalysts, Adsorbent materials	6.8 6.9

Pulse SP: Pulse-driven SP, DC SP: DC-driven SP, LPGD: Low-pressure glow discharge (dc or ac) on liquid, Arc SP: DC-driven arc SP, GDE: Glow-discharge electrolysis, MWSP: Microwave-driven SP

- with simple reactor
- with low energy consumption
- by one step (or one pot)
- at low temperature
- at atmospheric pressure
- within a short processing time

In short, SPP fits almost perfectly with the requirements for nanoparticle synthesis, and its fit led to a marked increase in the number of research activities on nanomaterial synthesis using SPPs.

2.1 *Bottom-Up or Top-Down*

Material synthesis, in general, is roughly categorized into the following two approaches, as shown in Fig. 1. The same classification can be applied to SPPs.

- **Bottom-up approach**:

This approach is based on the arrangement of material components. The typical size of the material components to be arranged is atomic or molecular level. The formation of nanomaterials involves nucleation and growth, which are caused by the collision of the material components. In the case of this approach, solvents, and stabilizing agents may influence the growth of nanomaterials.

- **Top-down approach**:

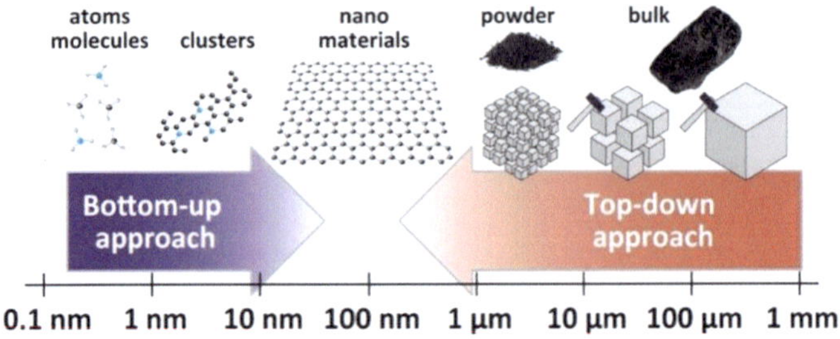

Fig. 1 Top-down versus bottom-up

This is based on the breakdown of macroscopic and bulky materials into nanoscale structures, which require external energy to break up and fragment the bulk material into smaller parts. Physical, chemical, thermal, or some other energy may be used for this approach.

The choice depends on the requirements of the desired material and/or synthesis process. Of course, these two approaches may be combined when necessary or unintentionally.

2.2 *Bottom-Up Approach*

In the case of SPPs based on the bottom-up approach, the origins of nanomaterials to be synthesized are the chemicals mixed in the solution. Typical SPPs are performed in an aqueous solution (Note that SPPs can also be performed in a non-aqueous organic solution as described later). The chemical species generated by the plasma in the aqueous solution involve both reductive and oxidative species (Šunka et al. 1999; Baroch et al. 2008; Potoký et al. 2009; Miron et al. 2010, 2011). Typical reductive and oxidative species are as follows;

- **Typical reductive species** (in the SPPs with the aqueous solution):
- H radicals, solvated electrons, H_2, and H_2O_2
- **Typical oxidative species** (in the SPPs with the aqueous solution): OH radicals and O atoms

The reductive species reduces metal ions to synthesize metal nanoparticles. Oxidative species produce oxides of transition metals or decompose organic compounds.

In the case of SPPs performed in the open air, care is necessary that the aqueous solution may be acidified due to a dissolution of N_2, O_2, and CO_2. The major cause of water acidification is the formation of NO_2 and HNO_3, which is caused by the reactions of plasma with the dissolved N_2 and O_2.

SP in organic solvent generates C_2, CH, and H radicals, which react in the organic solvent to form carbon materials such as polymers. In this case, it is necessary to avoid contamination of the liquid by the electrode material. Generally, tungsten (W), which has the highest melting point (3695 K) among metal materials, is used as the electrode. On the other hand, the top-down approach uses the electrode as a raw material, which is described below.

2.3 *Top-Down Approach*

The electrical discharge is usually ignited between a pair of electrodes. The top-down approach in SPPs employs the electrode as a source material for synthesizing nanomaterials. Since an electrode must have high electrical conductivity, metal or carbon is chosen as the electrode material. There are two possible conceptual pathways for the formation of nanomaterials in this approach.

- **Sputtering of an electrode**

The surface of an electrode during an electrical discharge is subjected to bombardments by high-energy ions, which causes the sputtering of the surface atoms of the electrode. The sputtered atoms work as the nuclei of the material to be synthesized and then transformed into nanoparticles through subsequent diffusion and growth in the solution.

- **Erosion of an electrode**

The current density becomes very high on the electrode surface during discharge. This makes the temperature of the electrode surface extremely high. If the temperature becomes high enough to melt the electrode surface, molten particles made of the electrode material are expelled from the electrode surface into the liquid. The liquid surrounding the injected molten particles cools down immediately and turns them into NPs made of the electrode material.

One of the advantages of these methods is that they do not need to add complex chemicals other than solvents (usually water). Of course, it can intentionally combine chemical processes between sputtered and/or eroded atoms with reactive species in the solution involving additional chemicals, which enables us to synthesize metal oxides and metal carbides.

3 Noble Metal Nanoparticles

According to a loose definition in chemistry, noble metals are as follows: silver (Ag), gold (Au), platinum (Pt), palladium (Pd), rhodium (Rh), ruthenium (Ru), iridium (Ir), and osmium (Os). Noble metals are less reactive to oxidation. In other words, their ions are readily reduced to form neutral atoms. Thus, the synthesis of noble metal NPs has generally been performed by using reduction reactions in the solutions, including metal salt precursors. Typical metal precursors are $HAuCl_4$, $AgNO_3$, H_2PtCl_6, and $PdCl_2$.

However, conventional reduction methods inevitably require a reducing agent and involve multiple processes, resulting in a time-consuming procedure as a whole. Therefore, in the synthesis of noble metal NPs, it is desirable to use as few chemical agents as possible. Of course, it also needs to be quick and easy. SPPs meet these demands because they are capable of metal-salt reduction within a short time without adding a reducing agent. The reduction reaction in the SPPs occurs in almost the same manner as the redox sequence seen in the electrochemical reaction.

The shape and size of metal NPs are critical characteristics because they strongly affect the optical and catalytic properties of the metal NPs. Extensive studies have been conducted to discover the key process parameters to control shape, size, and dispersity. The following subsections describe several important achievements of these studies.

3.1 Formation Mechanism

Saito et al. (2009), (Hieda et al. 2008a) studied the mechanisms of Au NPs synthesis in SPPs under various experimental conditions. They found that Higher voltage and higher initial concentration are effective in increasing the rate of reduction of Au ions and the formation of Au NPs.

During the synthesis of Au NPs using SPPs, NPs with unique shapes are formed depending on the condition, as shown in Fig. 2. For example, when the discharge time is very short, a dendrite-like structure is formed. They are loosely aggregated small Au nanoclusters. Although the size of the dendrite-like particles is as large as 100 nm or larger, with a longer discharge time of 45 min, the particle size decreases, and its shape becomes anisotropic, such as triangles, pentagons, and hexagons. The reduction in particle size caused by the prolonged discharge is explained by the decrease in the pH, because low-pH aqueous solutions have the ability to dissolve Au NPs.

3.2 Size Control

Hieda et al. (2008b) discovered a unique method to control the size of Au NPs. They used a reverse micelle solution, which is a mixture of sodium bis(2-ethylhexyl)sulfosuccinate, dodecane, and water). Reverse micelles are often referred to as nano-reactors because they play the role of capsules that hold extremely small volumes of an aqueous solution and spatially confine the progress of chemical reactions. Thus, the size of Au NPs can be controlled by adjusting the size of the nano-reactors, i.e., the inner pore size of reverse micelles. They have confirmed that Au ions are reduced only in the nano-reactors made of the reverse micelles and successfully reduced the size of Au NPs to as small as 4 nm. They also found that the homogeneity improves as the reverse micelle size decreases.

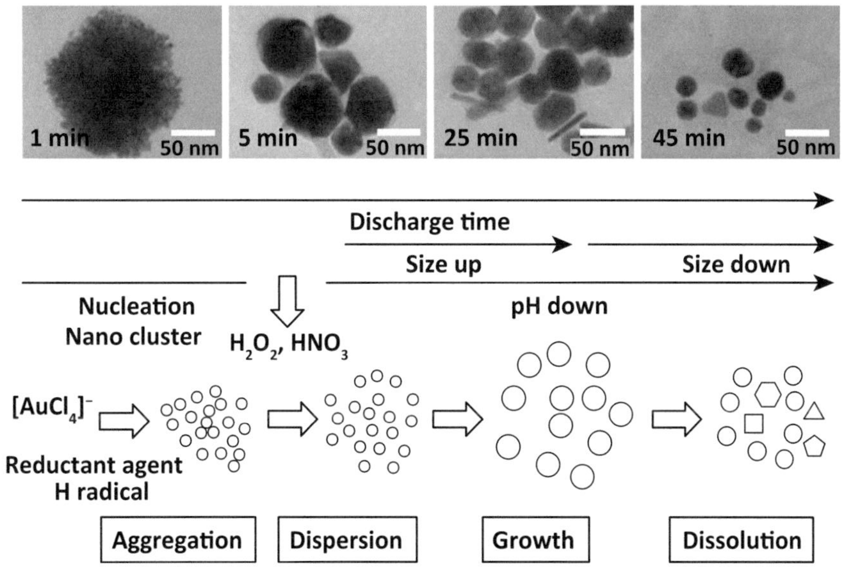

Fig. 2 The mechanism of Au NPs synthesis by SP in aqueous solution, and TEM images of Au NPs synthesized with different SP processing time (Saito et al. 2009)

Bratescu et al. (2011b) clarified the influence of pH on the size of Au NPs. They found that the size of Au NPs decreased from 10 to 1 nm when they increased the pH from 3 to 12. This tendency is explained in terms of the difference of the precursors for Au NP formation. Au ions in the aqueous solution exist in the form of $AuCl_4^-$ and $Au(OH)_4^-$ in the case of pH 3 and pH 12, respectively. They are both reduced by H radicals supplied from the plasma and turn into neutral Au, which works as the primary source of Au NPs. However, the standard redox potential (0.95 eV) for reducing $AuCl_4^-$ is higher than that (0.6 eV) for reducing $Au(OH)_4^-$. Since the higher redox potential causes supersaturation, $AuCl_4^-$ in the pH 3 solution tends to form larger Au NPs in comparison to $Au(OH)_4^-$ in the pH12 solution. In the case of the pH 12 solution, there is another factor for reducing the size of Au NPs. XPS analysis on the Au NPs prepared in the pH 12 solution revealed that the Au NP surfaces are partially oxidized and contain AuO^-, $AuOH^-$, and Au_2O^-. These negatively charged species on the Au NPs cause repulsive forces between them and prevent their agglomeration.

According to the report by Heo et al. (2014), we must pay attention to the temperature of the solution during Au NPs synthesis using SPPs. They found that lower temperatures lead to smaller nanoparticles and a narrower size distribution.

3.3 Reducing Agents in SPPs

Saito et al. (2009) proposed a mechanism for the reduction of Au ions to Au NPs. According to his reduction mechanism, the H radicals produced through the dissociation of H_2O can work as a reducing agent. This reduction process by H radicals in the SPPs has been evidenced by the results of electron spin resonance measurements.

Shirai et al. (2014) compared the formation mechanisms of Ag NPs from $AgNO_3$ and Au NPs from $HAuCl_4$. Their SPP was not performed in a typical SPP reactor. They used an H-shaped glass reactor to investigate the influence of electrode polarity. According to their experimental results, the Au NPs were formed near the anode, while Ag NPs were formed near the cathode.

This finding suggests that Ag and Au NPs are formed through different reduction mechanisms. Ag ions exist as Ag^+ in the solution. In contrast, major Au ions exist in the form of $[AuCl_x(OH)_{4-x}]^-$, where $x \leq 2$ at low pH and $x < 2$ at high pH. In this case, the Au and Ag ions move in different directions under the applied electric field in the solution.

They explained their results as follows: Ag ions are reduced by solvated electrons, while Au ions are reduced by H radicals. This mechanism also explains the SP synthesis of other noble metal NPs. For example, the reduction by H radicals can explain Pt NPs formation from H_2PtCl_6 aqueous solution (Ichin et al. 2009; Lee et al. 2014d). The reduction by solvated electrons can explain Pd NPs formation from $PdCl_2$ aqueous solution (Bera et al. 2004). Note that H-radicals are not the only agents of reduction. Other radicals may be involved if the aqueous solution has other solutes.

3.4 Enhancement of Production Rate

Sudare et al. (2015a, b) found that the production rate of Au NPs could be extremely enhanced (35-fold!) by adding ethanol to conventional water-based SPP conditions. The addition of methanol or 2-propanol showed a similar effect. According to their discussion, hydroxyethyl radicals (CH_3CHOH) play an important role in this extreme enhancement of the Au NP production process.

3.5 Microstructural Characteristics

Cho et al. (2011) revealed microstructural characteristics of the Au NPs using HRTEM. They prepared Au NPs of similar size using two methods for comparison; one was the SPP in reverse micelle solutions, and another was the conventional simple chemical reduction. The Au NPs fabricated by the SPP had the face-centered cubic (fcc) crystal structure with spherical morphology, multiply twinned particles (MTPs)

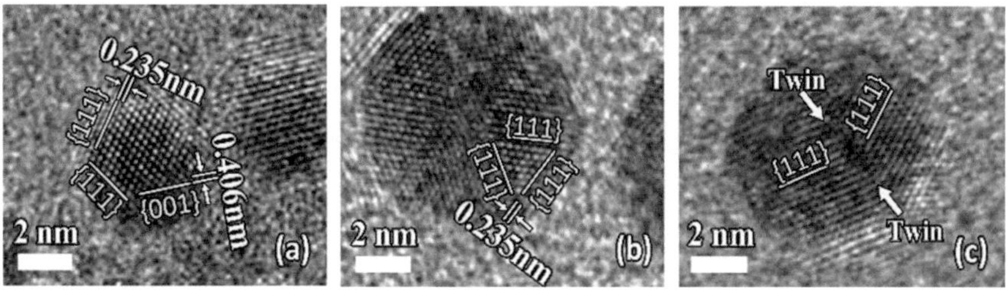

Fig. 3 HRTEM images of Au NPs synthesized by the SPP (Cho et al. 2011). **a** fcc single-crystalline particles with spherical morphology; **b** multiply twinned particles (MTPs) of icosahedral morphology with five-fold symmetry axes; **c** incomplete MTP (single-twinned fcc configuration)

of icosahedral morphology, and single nano-twinned fcc configurations (incomplete MTPs), as shown in Fig. 3.

The Au NPs synthesized using the conventional chemical reduction also had shapes and structures similar to those nanoparticles formed using the SPP. However, there was a difference in the fraction of MTPs contained in the synthesized Au NPs. In the case of conventional chemical reduction, the relative number of MTPs was 63%. In the case of the SPP, on the other hand, it was as high as 94%. Cho et al. explained this tendency in terms of the feature of the SPP, in which much more reactive species are involved in the reaction field than in conventional chemical reduction processes.

3.6 *Seed-Mediated Growth*

Ashkarran (2012) investigated the seed-mediated growth of Au. They found the formation of Au NPs during the post-discharge period, which was as long as hours and days. This result indicates that the Au NPs (or smaller clusters) created during the SPP have active surface sites which work as nucleation sites during the post-discharge period.

3.7 *Influence of Surfactants*

Conventional reductive synthesis of NPs employs a surfactant to prevent the aggregation of NPs. The same goes for SPPs. The following surfactants have been used to avoid the aggregation and maintain the dispersity of NPs:

- polyvinylpyrrolidone (PVP) (Kim et al. 2008)
- sodium dodecyl sulfate (SDS) (Saito et al. 2009; Hieda et al. 2008a)
- cetyltrimethylammonium bromide (CTAB) (Lee et al. 2013)
- cetyltrimethylammonium chloride (CTAC) (Bratescu et al. 2011b)

However, when we use noble metal NPs as catalysts, surfactant-free synthesis is essential. The surface of noble metal NPs involves active sites for catalytic reactions. If the surfactants mask the NP surfaces, we cannot expect effective catalytic reactions by the NPs. In an idealized situation, SPPs may avoid aggregation without surfactants. The reason is that the NPs in the SPP are generally negatively charged and repel each other. However, practical methods have not yet been established, and further studies are necessary.

3.8 Synthesis of Clusters

Nanoclusters (NCs) are defined as particles with a diameter of not more than 2 nm (within 150 atoms). The electronic structure of the particle is different from that of bulky particles (Zhu et al. 2021; Mingos 2014a,b). For example, the energy level of conduction band electrons changes from continuous to discrete. In addition, the proportion of atoms exposed on the surface exceeds 60%, and almost 50% of atoms are located at unsaturated corners and edges. Therefore, one can expect new physical properties not found in bulk and NPs.

In conventional chemical approaches, nanoclusters have been synthesized from metal ions or metal complexes in the presence of a reducing agent and stabilizing agents to inhibit the growth of larger NPs. However, such additional agents work as impurities, which should be removed by subsequent purification processes (Brust et al. 1994). One of the possible methods to overcome this issue is sputtering. In the sputtering, atoms are effused from the targets of pure materials. NPs created by sputtering, therefore, contain fewer impurities than those created using chemical methods. However, conventional sputtering performed in a vacuum system cannot synthesize size-controlled and high-dispersibility NCs with diameters less than 2 nm because of the weak interaction between the metal and dispersion medium (Schwyn et al. 1988; Torimoto et al. 2006).

Hu et al. (2012) fabricated well-crystalline clean gold clusters with spherical morphology (1.25 nm $\pm$ 0.5 nm) by using SPPs with Au electrodes, as shown in Figs. 4 and 5. Their method is categorized into the so-called top-down approach, in which Au electrodes for the SPP are used as the target materials in a sputtering process. Thus, this method is called solution plasma sputtering (SPS). Since energetic ion bombardment cannot be expected on the target in a highly collisional high-density medium, what is happening in this SPP is not physical sputtering but chemical sputtering (or erosion).

Note that SPS is conducted under cryogenic conditions using liquid nitrogen. This is an essential factor in suppressing the size of synthesized particles. The formation of NPs consists of sequential processes for nuclei formation and those for their growth. In these processes, the temperature has a greater influence than the concentration of the precursor. The reason is that the concentration contributes by the power law of the reaction order, while the temperature contributes exponentially. Thus, reduction of process temperature is an appropriate strategy for obtaining particles of smaller

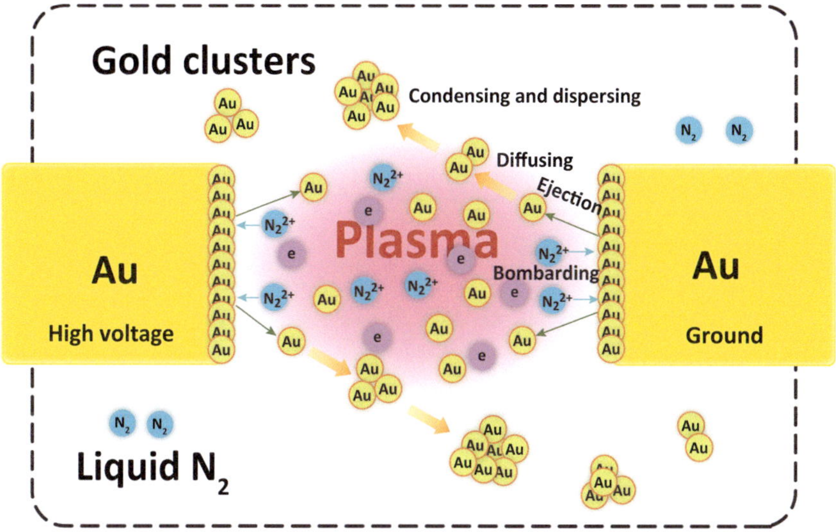

Fig. 4 Schematic illustration of the model for gold clusters in LN_2 by the solution plasma sputtering (Hu et al. 2012)

Fig. 5 A HRTEM image for a gold cluster synthesized by the top-down SPP (or SP sputtering) (Hu et al. 2012)

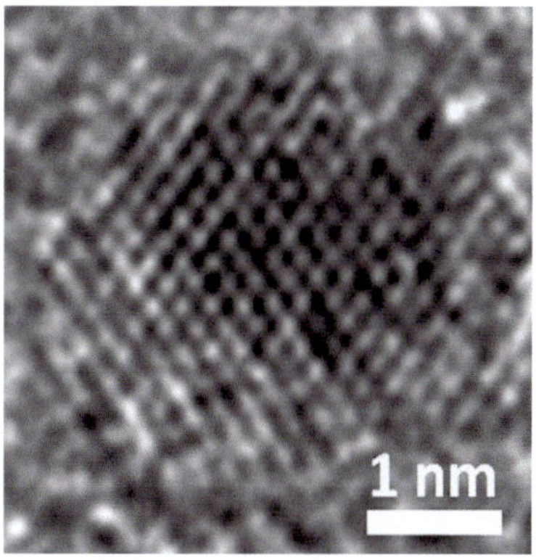

size. Conventional solution chemistry, however, requires thermal energy to proceed with chemical reactions to create nuclei. Thus, temperature reduction results in poor nuclei formation rate. Solution plasma sputtering, on the other hand, does not require thermal energy because plasma provides active species for sputtering to create nuclei, even if the surrounding medium is at cryogenic temperatures.

4 Non-Noble Metal Nanoparticles

SPPs can synthesize various non-noble metals. NPs as follows;

- Cu (Xie et al. 2004; Lo et al. 2005; Pootawang et al. 2013b)
- Sn (Lee et al. 2014b; Saito et al. 2014a,d; Kelgenbaeva et al. 2015)
- Fe (Kelgenbaeva et al. 2014)
- Co (Kim et al. 2013a)
- Ni (Saito et al. 2010)

In general, non-noble metal NPs are difficult to synthesize. The reason is that non-noble metals are readily oxidized when they meet oxidizing agents such as O and OH radicals, which are generated during SPP involving water. Even after the SPP, H_2O_2 molecules may exist in the aqueous solution and work as an oxidizing agent. This means that SPPs will produce oxides if no countermeasures are taken for oxidation. In the case of attempts to synthesize Cu NPs, Cu_2O and CuO will be formed (Xie et al. 2004; Pootawang et al. 2013b; Saito et al. 2011b).

Thus, we must add encapsulating agents and/or surfactants to the solution for immediate protection of emerging particles from oxidation. The validity of this strategy was proved in the synthesis of Cu NPs using SPPs, in which the following agents were used:

- CTAB + ascorbic acid (Xie et al. 2004)
- gelatin + ascorbic acid (Pootawang et al. 2013b) (Fig. 6)
- citrate buffer solutions (Saito et al. 2011b)

Ethylene glycol is also known to work in preventing the oxidation of Cu NPs (Lo et al. 2005). However, pure Cu NPs were formed only in pure ethylene glycol. In the case of a mixture with water, Cu_2O and CuO were formed.

Regarding the synthesis of Sn NPs, quite a lot of CTAB was necessary to prevent their oxidation in the early days of the research (Saito et al. 2014d; Lee et al. 2014b). After that, however, it was reported that Sn NPs could be synthesized without CTAB only by adjusting SPP conditions (Saito et al. 2014a).

Regarding the synthesis of Fe NPs, Kelgenbaeva et al. (2014) reported a unique method to prevent oxidation. They performed the synthesis of α-Fe NPs in the micro-emulsion solution, which is prepared by mixing toluene and water. The characteristic point of this process is that the synthesis of Fe NPs proceeds at the interface of water and toluene. Thus, synthesized Fe NPs are immediately covered with thin protective layers. Thanks to this situation, synthesized Fe NPs are isolated from both oxidation and/or carbonization.

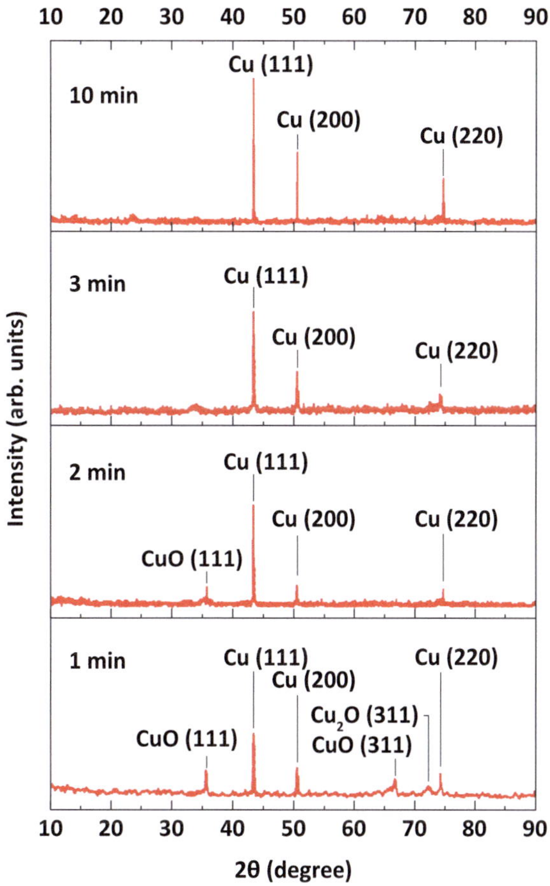

Fig. 6 XRD patterns of Cu NPs synthesized by an SPP at various discharge times of 1, 2, 3, and 10 min, in which oxide peaks disappear after 3-min synthesis (Pootawang et al. 2013b)

5 Bimetallic and Alloy Nanoparticles

A wide variety of bimetallic and alloy NPs can be synthesized by using SPPs. Hu et al. (2013) and Bratescu et al. (2013) successfully synthesized PtAu NPs by using the top-down SPP (or SP sputtering) in water without any reducing agents and stabilizers. Figure 7 shows the schematic illustration of this process, in which they used Au for one electrode and Pt for the other electrode. Note that the above works are based on the pioneering work by Terashima et al. (2013), although at that moment, what they obtained were not pure PtAu NPs but a mixture with Au NPs and Pt NPs. By changing the electrode pair from Pt-Au to Pd-Au, PdAu NPs can also be synthesized (Hu et al. 2015). It should be noted that the bimetallic and alloy NPs can also be synthesized in organic solvents such as benzene (Kang and Saito 2015).

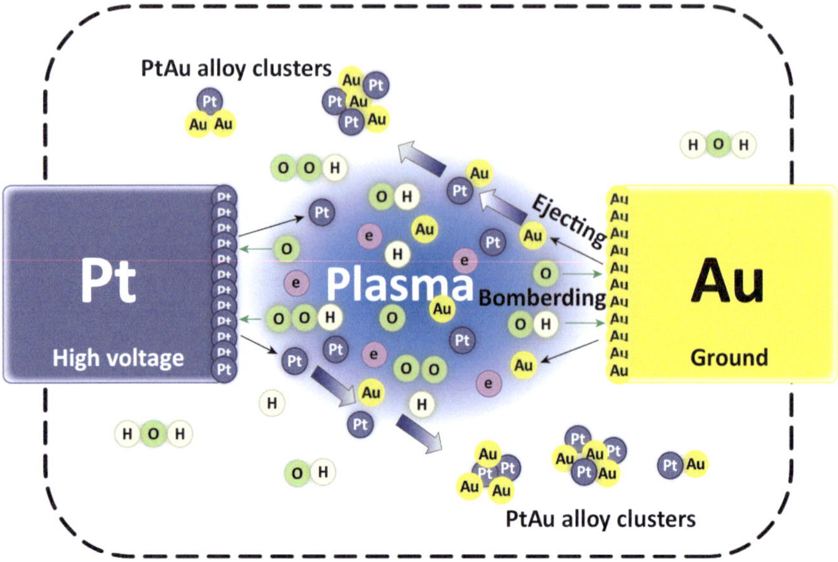

Fig. 7 The model of synthesizing PtAu alloy clusters in water by SP sputtering (Hu et al. 2013)

5.1 Influence of Electrode Gap

Electrode gap has a significant influence on the discharge characteristics in the SPP and also on the synthesized NP characteristics. According to the research by Hu et al. (2013) and Bratescu et al. (2013), the gap between Au and Pt electrodes should be as small as 0.3 mm for getting PtAu NPs. A wider electrode gap causes the formation of an un-overlapped plasma region, which results in the separated formation of Au and Pt clusters. When the PtAu NPs were synthesized under the optimal conditions, their diameter was as small as 2 nm. In addition, their aggregation was almost completely suppressed in water (also in ethanol) without surfactant addition.

5.2 Influence of Power Supply

In the case of the synthesis of bimetallic or alloy NPs, the characteristics of a power supply also have an influence on the synthesis processes. All studies described above employed a power supply that provides bipolar pulse waveforms. The reason is that if the polarity of the pulses is symmetric, we can expect the same sputtering rate on both metal electrodes. What happens in the case of a non-symmetric pulsed power supply? Pootawang et al. (2012b) investigated the synthesis of PtAg NPs using a unipolar pulse power supply, in which they used Ag for the cathode electrode and Pt for the anode electrode. What they obtained was not complete PtAg alloy NPs, but bimetallic composites, which were composed of 95% of Ag and 5% of Pt.

What happens when a DC power supply is used instead of a bipolar power supply? Kim et al. (2014d) studied to answer this question. They prepared PtPd NPs using DC and bipolar power supplies. In the case of the DC power supply, the synthesized products were mainly composed of cathode-side metal. In the case of the bipolar power supply, the synthesized products contain pure metal (Pt or Pd) NPs together with PtPd alloy NPs. Another important finding was reported by Kim et al. (2015b). They studied the influence of pulse duration (or duty) of the bipolar pulse waveforms on the synthesis of PtAg NPs. They found that pulse duration is a very important parameter for controlling alloy compositions. These results mean that the optimization of SP sputtering processes requires study from the viewpoints of not only chemistry but also electrical and electromagnetic engineering.

5.3 Bimetallic Au NPs with Various Other Metals

Bratescu et al. (2013, Bratescu and Saito 2013), (Saito et al. 2014b) synthesized bimet al. lic NPs by combining two approaches of top-down (i.e., reduction of ions) and bottom-up (sputtering of electrodes) in the one-step SPP. They synthesized various bimetallic Au-X NPs using Au electrode and nitrate of the metal X, where X was Zn, Cd, Ga, In, Fe, Co, Ni, Cu, Ag, and Pd. Although this unique synthesis method employing SP sputtering and reduction has the possibility to synthesize various alloys, further study is required to control the size and composition. For example, the synthesized bimetallic Au NPs showed a strong tendency to aggregate. It should also be noted that the Au fraction becomes very high (80 wt.% or more) when the synthesis was attempted with Fe, Co, Ni, and Cu, which are categorized in 3d metals, in comparison to the other cases.

6 Metal Oxide

Synthesis of pure non-noble metal NPs is a difficult task, as described before because they are easily oxidized. This also means that the synthesis of their oxides is easy. The SPPs using an aqueous solution generally involve oxidizing agents known as reactive oxygen species (ROS). The representatives ROS in the SPPs are superoxide ($O^{-}{}_{2}$) and hydroxyl (·OH) radicals, which are produced through the reactions between plasma and water molecules.

Metal oxides in solution plasma sputtering are formed via sputtering or eroding of metal atoms from metallic electrodes and their subsequent oxidation by ROS. The following metal oxides have been synthesized using solution plasma sputtering in water:

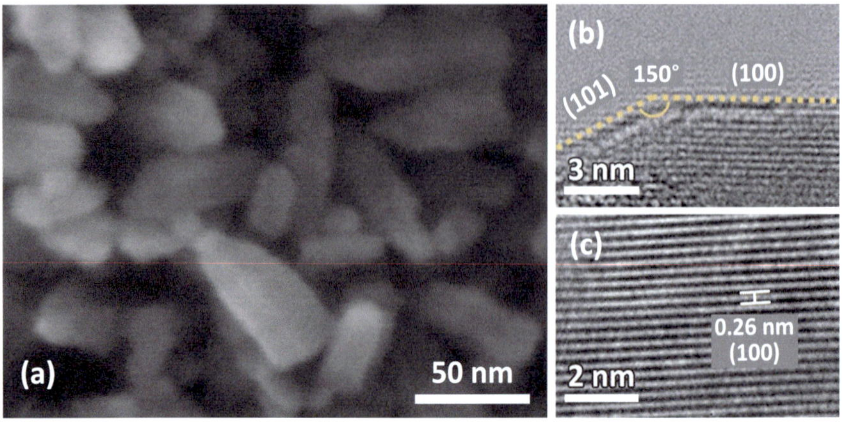

Fig. 8 ZnO nano-bullets prepared by the top-down SPP (or SP sputtering) (Kim et al. 2021)

- ZnO (Ashkarran et al. 2009b, 2010c; Kim et al. 2021)
- ZrO_2 (Ashkarran et al. 2010b)
- TiO_{2-x} (Ashkarran et al. 2010a; Panomsuwan et al. 2015d; Jedsukontorn et al. 2017, 2018)
- WO_3 (Ashkarran et al. 2008; Ashkarran 2011)

A common tendency for them was that their size increased with increasing discharge current. Another feature of the metal oxides was that they have a strong tendency to have a spherical shape. The reason for this tendency is explained in terms of the relatively high surface tension of these oxide materials.

In contrast, ZnO particles exhibit a unique feature. Their shape tends to be rod-like (Ashkarran et al. 2010c). The tendency is explained in terms of rapid cooling. Thus, they also change into spherical if the process temperature significantly increases by increasing the discharge current.

Kim et al. (2021) discovered that ZnO nano-bullets could be synthesized using solution plasma sputtering, as shown in Fig. 8. The main exposed facet of the ZnO nano-bullets was the (100) facet, while the (101) facet was confirmed on the inclined surfaces in their edges. The (101) facet has an asymmetric Zn and O atom arrangement with higher electron density, which causes higher reactivity. Thus, the ZnO nano-bullets are promising materials for application in advanced technology such as green chemistry.

6.1 *Oxygen Deficiency*

Note that an undesirable phenomenon occurs if top-down synthesis of the metal oxide is performed in water, where the oxygen is deficient. Oxygen deficiency results in insufficient oxidation of metals. Thus, synthesized products tend to be mixtures composed of pure metal and metal oxide, as observed when the SP sputtering was performed with a Ti electrode (Ashkarran et al. 2010a; Zhang et al. 2011; Nakasugi et al. 2014; Panomsuwan et al. 2015d), in which synthesized particles were composed of Ti, TiO, Ti_2O_3, and TiO_2. This tendency was also reported when the metal element was Zn (Ashkarran et al. 2009b, 2010c; Burakov et al. 2008; Tarasenko et al. 2010) and Sn (Kelgenbaeva et al. 2015). Several attempts to overcome this issue were proposed as follows;

- O_2 bubbling (Burakov et al. 2008)
- H_2O_2 addition (Ashkarran et al. 2010a)
- thermal treatment in the air (Kelgenbaeva et al. 2015; Ashkarran et al. 2010c; Ashkarran 2011)
- post-synthesis aging (Burakov et al. 2008)

Significant improvement in the oxidation level can be achieved by these methods, although there have been difficulties in some cases.

The other methods were also reported. In the case of the synthesis of ZrO_2 NPs, the addition of ammonia was reported to be effective in avoiding oxygen deficiency (Chen et al. 2011b). This method was also effective in improving crystallinity. The combination of DC plasma and the mixing of H_2O_2 is also known to be effective (Ashkarran et al. 2010b). In the case of the synthesis of magnetite (Fe_3O_4), the mixing of cetylpyridinium bromide (CPyB) is reported to be effective in suppressing the formation of pure Fe and incompletely oxidized FeO (Kelgenbaeva et al. 2013).

6.2 Exotic Shape

Metal oxide NPs were formed as exotic-shaped NPs such as nanoflowers, square plates, and nanofibers, which were synthesized under specific controlled conditions and environments.

Hu et al. (2014) found ZnO nanoflowers by solution plasma sputtering with Zn electrodes in H_2O. Figure 9 shows a STEM dark-field image of the synthesized ZnO nanoflowers. They revealed that the ZnO nanoflowers had good crystallinity from the results of HRTEM observation, and they grew along the 001 direction. They explained the formation mechanism of the ZnO nanoflowers as follows: The first step is the oxidization of Zn atoms. This results in the formation of ZnO nanoclusters, which work as nuclei for further growth of them. In this growth phase, anisotropic growth, which causes the exotic shape, proceeds because of the difference in the surface energy of each crystal face of ZnO. The (001) face of the ZnO is polar with high surface energy, and its thermodynamical stability is relatively low. On the other hand, the (100) face of the ZnO is non-polar with low surface energy and relatively stable. Because of such differences, the rearrangement of atoms on the (001) face occurs more easily than on the (100) face.

Saito et al. also studied the synthesis of nanoflowers of ZnO (Saito et al. 2011a, 2014c) and also CuO (Saito et al. 2011b) by using SPPs. According to their research,

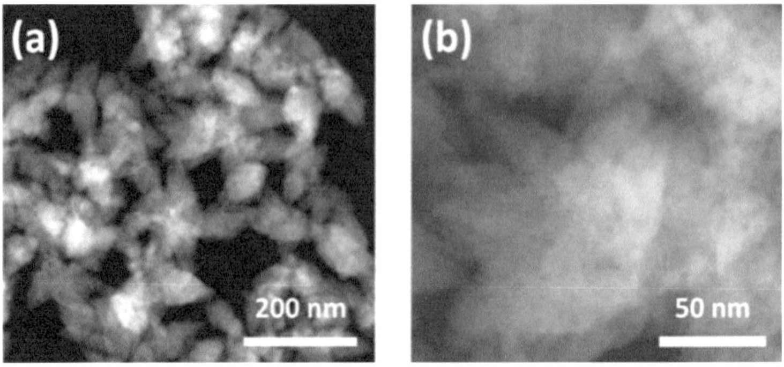

Fig. 9 ZnO nanoflowers prepared by solution plasma sputtering (Hu et al. 2014)

ZnO nanoflowers tend to be synthesized when the temperature of SPP is very high. They also investigated the influence of stirring during their SPP. They found that synthesized materials tend to be amorphous and/or oxygen-de ficient materials when stirring is employed. These results suggest that the process window for the synthesis of nanoflowers is very narrow.

Another exotic shape is nanofibers, which were found in the products of the SPP for the synthesis of CuO particles, as shown Fig. 10 (Xie et al. 2004; Hu et al. 2014). The process window for synthesizing CuO nanofibers was reported to be wider than that of ZnO nanoflowers. They also reported various exotic shapes found in the case of SnO synthesis depending on the process parameters and additive chemicals (Saito et al. 2012).

Fig. 10 CuO nanofibers prepared by solution plasma (Hu et al. 2014)

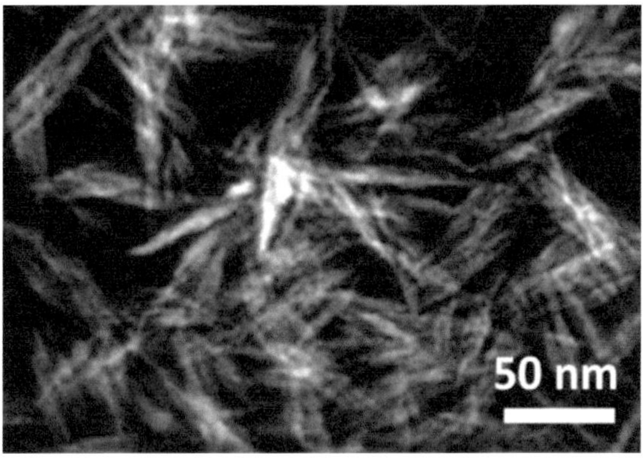

6.3 *Reduction then Oxidation*

Meta oxides are also synthesized by a bottom-up approach, in which dissolved metal ions are once reduced by H radicals to form NPs, and then oxidized by ROS such as OH and O in the solution to form meta oxides. In the case of this kind of multi-step reaction, identification of the rate-limiting step is important for process control. In the case of the SPP using $FeCl_3$, oxidation is faster than reduction, and FeO and FeO(OH) are produced (Bratescu et al. 2011a).

In the case of the SPP using $ZnCl_2$, oxidation to synthesize ZnO occurs only when urea is mixed in the solution (Tong et al. 2012). In the case of the SPP using $FeCl_3$ and NaF, the addition of urea causes the formation of mesoporous amorphous FeOF, which has the shape of a cocoon (Fu et al. 2015). In the case of the SPP using titanium bis (ammonium lactate) dihydroxide, TiO_2, having the shape like leaves, is synthesized only when Brij-30 is mixed into the solution (He et al. 2014), where Brij-30 is polyoxyethylene(4) lauryl ether (Lipshutz and Ghorai 2008). As can be understood from these reports, mixing additional chemical agents to the SPP brings about the synthesis of metal oxides with unique shapes. Synthesis of metal oxides via the reduction and oxidation of metal ions were also reported for the following materials:

- CoO (Chen et al. 2012)
- NiO (Wu et al. 2012)
- SnO_2 (Zhang et al. 2015)
- MnO_2 (Kim et al. 2016)

7 Carbon Materials

7.1 *Top-Down Approach*

The synthesis of carbon materials by SPPs has attracted much attention since SPPs are free from many constraints involved in conventional methods such as chemical vapor deposition, arc discharge, pyrolysis, and combustion. These conventional methods require, for example, complex equipment with a high-cost vacuum system and operation at extremely high temperatures. On the other hand, SPPs can be performed with very simple equipment at room temperature under atmospheric pressure conditions without any vacuum systems.

In the case of SPPs, you can synthesize carbon materials by either both top-down or bottom-up approaches (Li et al. 2018). In the case of a top-down approach, you can use a graphite electrode as a carbon source. In the case of a bottom-up approach, you can use liquid organic compounds as a carbon source.

Sano et al. (2001, 2002) proposed submerged high-temperature arc discharge in water, in which graphite electrodes are used as carbon sources. In this method, carbon atoms are effused (vaporized) from the graphite electrodes, and then carbon nanomaterials are synthesized through their reactions and rapid cooling processes. One of the features of this method is the unique cross-sectional structure of synthesized carbon nanomaterials, which have an onion-like structure. They call them carbon nano-onions. This method can synthesize also carbon nanotubes (CNTs) when ionic metal salts are mixed into water (Zhu et al. 2002; Lange et al. 2003).

Lee et al. (2014a) suggested the importance of input power for controlling the availability of carbon nano-onions. They performed the synthesis of carbon materials under different power inputs. When the input power is high enough, carbon nano-onions are synthesized, as shown in Fig. 11. However, if the input power is low, synthesized carbon materials become graphene flakes, as shown in Fig. 12, which are simply exfoliated from the carbon electrodes.

Omurzak et al. (2015) suggested the importance of the pulse duration or duty ratio between on and off. According to their experiments, if the duty ratio becomes larger, the hollow size of the nano-onions becomes larger, and the number of shells becomes smaller. In addition to these findings, they also disco vered the liquid medium dependence. According to their explanation, the growth of shells becomes significantly faster when they use a medium with a higher number of carbon atoms. Namely, the growth in toluene is faster than in ethanol.

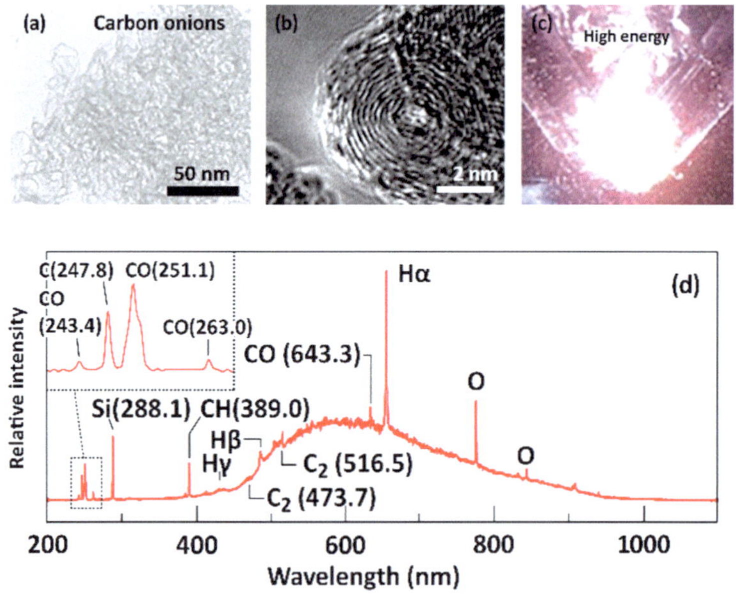

Fig. 11 **a** TEM and **b** HRTEM images indicating formation of carbon onions by the SPP at high-energy condition as confirmed from **c** the high-intensity discarge and **d** the OES spectrum (Lee et al. 2014a; Lee 2016)

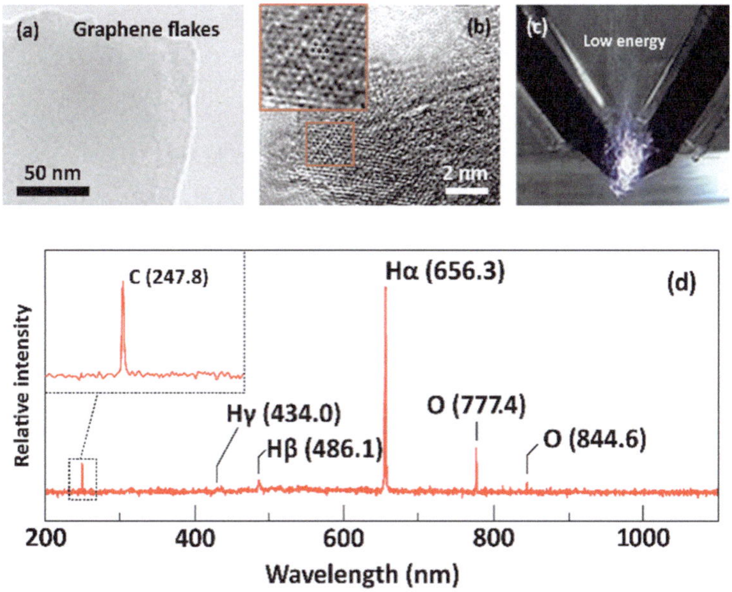

Fig. 12 **a** TEM and **b** HTREM images indicating formation of graphene flakes by the SPP at low-energy condition as confirmed from **c** the low-intensity discharge and **d** the OES spectrum (Lee et al. 2014a; Lee 2016)

7.2 Bottom-Up Approach

The bottom-up approach for the SPP synthesis of carbon nanomaterials is generally conducted by using organic hydrocarbons. There are many kinds of hydrocarbon materials that can be used for the SPPs as follows:

- benzene (Li et al. 2013a,b; Kang et al. 2013a,c; Yun et al. 2013, 2014a,b,c; Lee et al. 2015b)
- hexane (Kroushawi et al. 2012)
- dodecane (Fisher and Thagard 2012)
- alcohol (Hagino et al. 2012; Matsushima et al. 2013)
- acetone (Fisher and Thagard 2012)

Among them, benzene is frequently used for the synthesis of carbon nanomaterials because it brings about a considerably fast production rate of 10–20 mg/min. An example of the synthesis of CNPs by an SPP using benzene is shown in Fig. 13a. According to the report by Kang et al. (Kang et al. 2013a, c), CNPs synthesized from benzene have the common tendency to have spherical or quasi-spherical shapes with a diameter of 20–40 nm. The CNPs from benzene also have a tendency to aggregate each other, which results in the formation of porous structures. Although the CNPs from benzene have crystallographic structure mostly dominated by amorphous phase, their assemblies with porous structure work very well when they are used for the

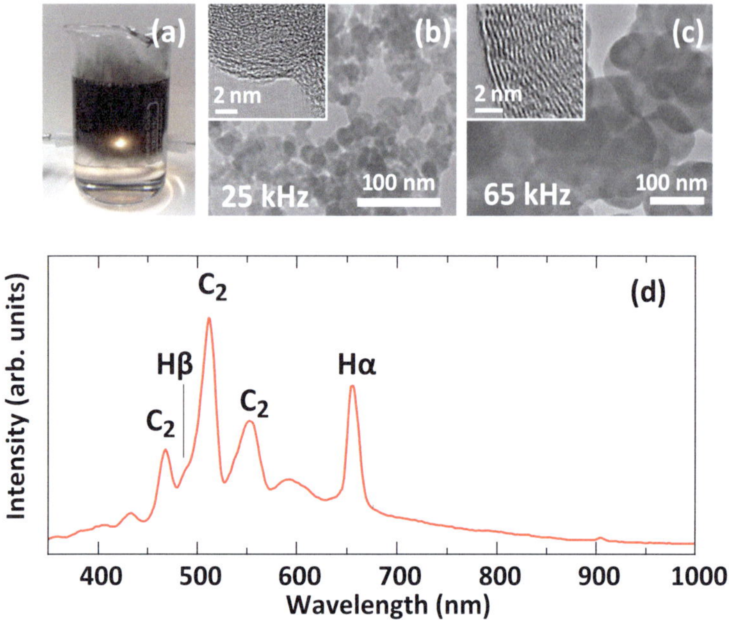

Fig. 13 **a** Overview of the reactor during the SPP in benzene. TEM images of CNPs synthesized from benzene with the frequency of **b** 25 kHz and **c** 65 kHz (Kang et al. 2013c). **d** An optical emission spectrum of the SP in benzene (Morishita et al. 2016)

electrodes of Li-air batteries because of enhancement in the discharge capacity owing to the mesoporous structures, although thermal annealing processes are required to improve their surface area and electrical conductivity (Yun et al. 2013).

In the case of conventional chemical synthesis, thermal annealing methods are routinely used to improve the characteristics of CNPs. In the case of SPPs, there are other available options, such as power, frequency, and duty ratio of pulse power. Among them, Kang et al. (2013c) and Hyun et al. (Hyun et al. 2016c) investigated the influence of pulse frequency, as shown in Fig. 13b–c. As can be understood from these figures, an increase in the pulse frequency brings about significant improvement of the CNP characteristics such as graphitization degree and electrical conductivity, although it brings about also increase in the particle size and decreases in the surface area.

In addition to the voltage waveform, electrode configuration also has a significant influence on some characteristics of CNPs. For example, the crystallinity and hydrogen content of the CNPs show electrode-gap dependence. According to the report by Lee et al. (2015b), the electrode gap should not be too narrow or too wide. They found 0.75 mm is the best condition.

Additional chemicals also work for improvement of synthesized CNP characteristics. Li et al. (2016) found that CCl_4 addition improves the electrical conductivity of synthesized CNPs. According to their discussion, Cl radicals scavenge excess hydrogen, which reduces electrical conductivity, by forming HCl. Although mechanisms should be different, the addition of naphthalene or anthracene has a similar effect. The addition of water also has favorable influences on the CNP characteristics, such as improvement in the crystallinity (Yun et al. 2014a,b,c).

The fundamental mechanisms of CNP formation by SPPs in benzene has been discussed by many researchers (Kang et al. 2013b; Yun et al. 2014c; Li et al. 2013a; Morishita et al. 2016). Figure 13d shows the OES of the SPP in benzene, in which you can see optical emission from C_2, CH, and H. This means that benzene molecules are fragmented into small species in the benzene plasma formed between electrodes. Among these species, C_2 is believed to be the source for producing CNPs.

In the case of SPPs, reactions at the gas/liquid interfacial region often play important roles rather than those in the central plasma region. This is due to the fact that the volume of the interfacial region is much larger than that of the central plasma region. Since the interfacial region is apart from the high-temperature plasma region, the temperature of the interfacial region is relatively low. In addition, polymerization is generally faster than carbonization at low temperatures. Because of this, polymerization reactions tend to proceed rather than carbonization at the interfacial region. Thus, major reactions to form CNPs in the SPPs in benzene are believed to be polymerization. Very high synthesis rates of the CNPs in the case of benzene are generally explained in terms of the fact that benzene molecules have unsaturated $C = C$ bonds, which boost the polymerization rates. As can be understood from these mechanisms, the opposite occurs when using hydrocarbons without $C = C$ bonds. For example, CNP formation in the SPPs in aliphatic hydrocarbons, which do not have $C = C$ bonds, occurs with very low production rates.

7.3 Heteroatom-Doped Carbon

Heteroatom doping of CNPs is one of the promising methods to enhance catalytic activities, or oxygen reduction reaction (ORR), of CNPs used for the cathode of fuel cells or metal-air batteries. The doping of CNPs is realized by the substitutional replacement of C atoms in the CNPs to desired dopant atoms. In the case of CNPs, dopants can be N, B, P, and F. Conventional doping is performed on non-doped CNPs that have already been synthesized. In this method, however, doping of heteroatoms to CNPs requires the removal of carbon atoms to be substituted with dopant atoms. This means that the carbon network should be destroyed once before the substitutional doping. If the substitution by the dopants is not fully accomplished, defects are left in the CNPs, which results in the reduction of catalytic activities of CNPs. In the case of SPPs, heteroatom doping is achieved simply by using heteroatom-containing source materials. One of the great advantages of this method is the fact that the synthesis and doping proceed simultaneously. This means that you don't need to destroy the carbon net work for the substitutional doping of heteroatoms to CNPs. Thus, we can expect heteroatom-doped CNPs with fewer defects by employing SPPs.

Synthesis of N-doped CNPs (N-CNPs) is one the most successful example of heteroatom-doped CNPs by SPPs. Panomsuwan et al. (2014) used pyrazine as a N-containing source material, which was used together with conventional benzenen. Kim et al. (2014b) used phthalocyanine as an N-containing source material together with benzene and dodecane. As described before, in the formation of CNPs by SPPs, C_2 radicals generated in the plasma are believed to play an important role in synthesizing CNPs. In the case of the synthesis of N-CNPs, it is reported that CN radicals are also important key radicals for N-doping (Panomsuwan et al. 2006, 2014).

Although N-CNPs can be synthesized by SPPs using a mixture of two source materials for supplying C and N, there is a difficulty that we cannot freely adjust the N concentration due to the solubility limit. Thus, it is better to employ a single source material that works as the source for the formation of a carbon network and also as a dopant. The following materials are typical examples;

- acrylonitrile (Panomsuwan et al. 2006)
- pyrrole (Kim et al. 2014a)
- pyridine (Thongwichit et al. 2016; Kim et al. 2015a)
- aniline (Hyun et al. 2016c)

Panomsuwan et al. (2016a, b) examined several source materials for preparing the N-CNPs. They deduced a simple relationship between the N-doping level in the N-CNPs and the N/C ratio in the source molecules. They found that the higher the N/C ratio in the source molecules, the higher the N-doping level in the synthesized N-CNPs. According to their experimental results, N-doping level in the synthesized N-CNPs increases in the order of benzonitrile (C_7H_5N), cyanopyridine ($C_6H_4N_2$), and cyanopyrazine ($C_5H_3N_3$). However, the rate of producing N-CNPs decreases in this order because the number of carbon atoms decreases in this order.

The catalytic activity (or the ORR rate) of N-CNPs strongly depends also on how (or where) the doped N atoms are incorporated into the CNP network. The favorable N-incorporation sites are reported to be graphitic and pyridinic sites (Kim et al. 2014b, Panomsuwan et al. 2016a, b; 2015b). Regrettably, most of the N atoms in as-synthesized CNPs are not in the graphitic and/or pyridinic sites. They are incorporated as nitrile, amine, or amino groups. One of the possible solutions to overcome this issue is thermal annealing, which has been known to improve the fraction of the N atoms at graphitic and/or pyridinic sites. However, this treatment is also known to be accompanied by a reduction in N content in the N-CNPs. Thus, further study is necessary for obtaining the N-CNPs having desired N-incorporation schemes.

It should be noted that the shape of synthesized N-doped carbon nanomaterials is not only particle shape. Hyun et al. (2016a, b) demonstrated the synthesis of N-CNPs having sheet-like structures. They found such unique structures when they used N-methyl pyrrolidone (NMP) as source material, as shown in Fig. 14. It should also be noted that the sheet-like structures were available only when they used NMP as source material and employed a relatively high pulse frequency of 200 kHz. When they used conventional benzene and aniline as source materials, they could not obtain sheet-like structures but particles, even if they employed a frequency of 200 kHz.

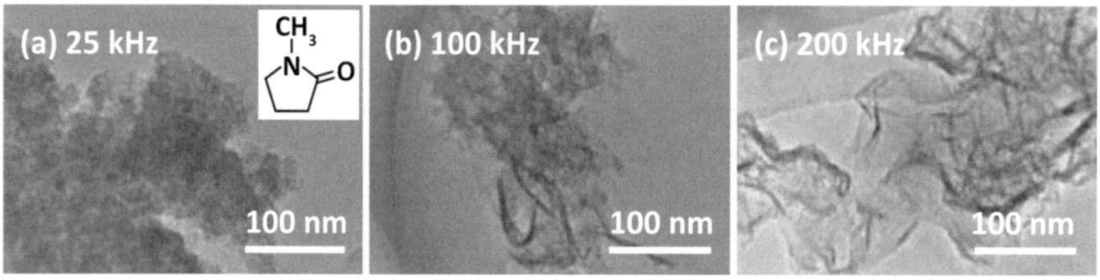

Fig. 14 TEM images of N-doped carbon nanosheets synthesized by the SPP using NMP as a source materrial at different frequencies of **a** 25 kHz, **b** 100 kHz, and **c** 200 kHz (Hyun et al. 2016a)

At present, it has not yet clarified why such non-spherical shapes are synthesized only under specific conditions and source material. By understanding its mechanism, however, further development of SPP technologies is expected.

7.4 Beyond N-doped Carbon

Doping with atoms other than nitrogen has also been investigated to improve the catalytic activity of carbon nanomaterials. For the same purpose, surface modification of carbon nanomaterials by functional groups has also been attempted.

F-doped CNPs (F-CNPs) were reported by Panomsuwan et al. (2015c). They used toluene and trifluorotoluene as the source materials. According to their investigation, the doped F atoms in the synthesized F-CNPs are incorporated into the carbon network by forming ionic bonds, which can promote catalytic activity such as ORR.

B-doped CNPs (B-CNPs) were reported by Panomsuwan et al. (2015a). They used triphenylborate ($C_{18}H_{15}BO_3$) together with benzene. They achieved not only higher catalytic activity but also longer durability and stronger methanol tolerance than conventional CNPs.

Doping with multiple different atoms has also been attempted. Kim et al. (2015a) reported doping with B and N and with P and N. They used benzene and pyridine as the source materials for synthesizing N-CNPs. Co-doping with B and P was performed by mixing phenylboronic acid ($C_6H_7BO_2$) and phenylphosphonic acid ($C_6H_7PO_3$), respectively. They found that the CNPs prepared by these co-doping processes showed significantly enhanced catalytic activity in comparison to that of singly doped N-CNPs.

The catalytic activity of CNPs is known to be improved by the modification of CNP surfaces with oxygen-containing functional groups, which was revealed by Ishizaki et al. (2014). They added 1,4 dioxane to benzene when synthesizing CNPs by SPPs, which resulted in the formation of COOH and/or OH on the surface of CNPs and enhancement of catalytic activity for the ORR.

In some cases, doping with heteroatoms or surface modification of CNPs may produce effects other than improving catalytic activity. Li et al. (2019) synthesized S-doped carbon nanoparticles using benzene and thioanisole as carbon and sulfur sources, respectively. They loaded Pt NPs onto the S-doped CNPs. From the sintering test at 600 °C, they found that S-doped CNPs showed superior Pt-support durability compared to pristine carbon matrix. This result suggested that the sulfate C-S-C bonds provided a strong interaction between Pt and S-doped CNPs. This means that S-doped CNPs can serve as a stabilizer of Pt NPs to enhance their durability in the application of metal-air batteries and other electrochemical devices.

7.5 Cationic N-doped Carbon

It has been known that the properties of N-doped graphene are strongly influenced by where N atoms are incorporated in the graphene network (Yadav and Dixit 2017). Possible N-bonding configurations, which are known as cationic N, graphitic N, and quaternary N, are shown in Fig. 15.

N atoms in the configuration of the graphitic N work as the donors in semiconductors. The excess valence electrons provided by these donors are delocalized and have high mobility. Thus, graphene with graphitic N has the electronic properties of an n-type semiconductor (Schiros et al. 2012). On the other hand, in the case of N atoms at the pyridinic N or pyrrolic N sites, they can be either donors or acceptors depending on the bonding configuration surrounding these sites (Guo et al. 2013). Therefore, valence electron control by pyridinic N or pyrrolic N is difficult. In other words, it is difficult to predict whether this type of graphene will be a p-type or n-type semiconductor.

However, when the N atoms have the bonding configuration described above, the surrounding chemical bonds cannot maintain a planar two-dimensional structure (Chai et al. 2014). This situation becomes worse when the number of dopants (N) increases. Thus, in some cases, we need the N bonding configuration without distorting the 2D graphene structure. According to Xu et al. (2017), it can be possible that a 2D planar graphene structure can be maintained by employing the bonding configuration called cationic N.

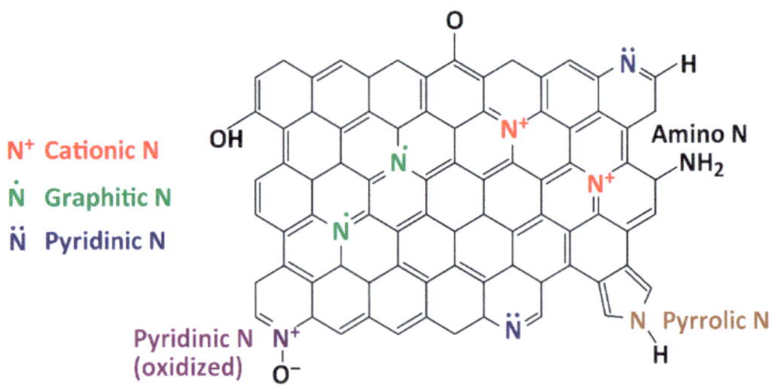

Fig. 15 Possible N-bonding configurations in the carbon network (Chae et al. 2020)

The 2D planar hexagonal structure of graphene is formed by the sp^2 hybridization of four valence electrons of a carbon atom. This situation is possible in the case of graphene because the nearest neighbor atom is also carbon. If the carbon is replaced by the heteroatom with a different number of valence electrons, the sp2 planar structure cannot be maintained. In the case of N-doping, we encounter this unfavoured situation because the N atom has five valence electrons. However, what if we dope ionized N^+ instead of neutral N? Since N^+ has one less valence electron than neutral N, it has four valence electrons, just like carbon. Thus, we can expect to be able to keep the planar structure even if we dope N^+. A similar strategy can be applied to doping B, which has one less valence electron. In other words, if we can dope B^- instead of B, we can maintain the flatness of graphene even after doping.

Chae et al. (2017) studied possibility of synthesizing cationic-N doped carbon (CN-C) or cationic-N doped graphene (CN-G) by SPPs. In their SPPs, they used graphite as the electrodes, which resulted in the formation of graphene sheets by peeling the electrode surface. Regarding the liquid medium, they used an aqueous solution of 1-ethyl-3-methylimidazolium dicyanamide (EMIM-DCA, $C_8H_{11}N_5$), which is a kind of ionic liquid. By this method, they successfully obtained a few-layered CN-G, which were confirmed by XRD, Raman spectroscopy, TEM, FE-SEM, FTIR spectroscopy, and XPS.

Chae et al. (2019) also revealed that it is possible to synthesize CN-G by using an organic solvent as the liquid, even when using a metal electrode instead of a graphite electrode, as shown in Fig. 16. In this case, they used dimethylformamide (DMF, C_3H_7NO) together with EMIM-DCA as source materials. The formation of CN-G by using this method was confirmed through characterization with TEM, Raman spectroscopy, and XPS. They also performed electrical measurements on the synthesized CN-G and revealed that the CN-G showed characteristics of a p-type semiconductor.

Chae et al. (2020) also synthesized CN-C and CN-C wrapped single-wall carbon nanotubes (CN-C@SWCNT) using the SPP shown in Fig. 17. The formation of these unique nanomaterials was confirmed through characterization with Raman spectroscopy, electron diffraction, and XPS. A schematic illustration of the CN-C@SWCNT is shown in Fig. 18. Electrical measurements were also performed on the CN-C@SWCNT. They revealed that synthesized CN-C@SWCNT shows p-type semiconductor characteristics with a high hole concentration of 4.6×10^{20} cm^{-3} and high electrical conductivity of 120 S cm^{-1}.

8 Pt-Free Catalysts for Oxygen Reduction

At present, Pt or Pt supported on carbon nanomaterials are believed to be the best material for achieving efficient catalytic activity for ORR shown in Fig. 19, which are employed in fuel cells and metal-air batteries. However, due to its high cost and poor methanol tolerance, it has not yet been put to large-scale practical uses. Thus,

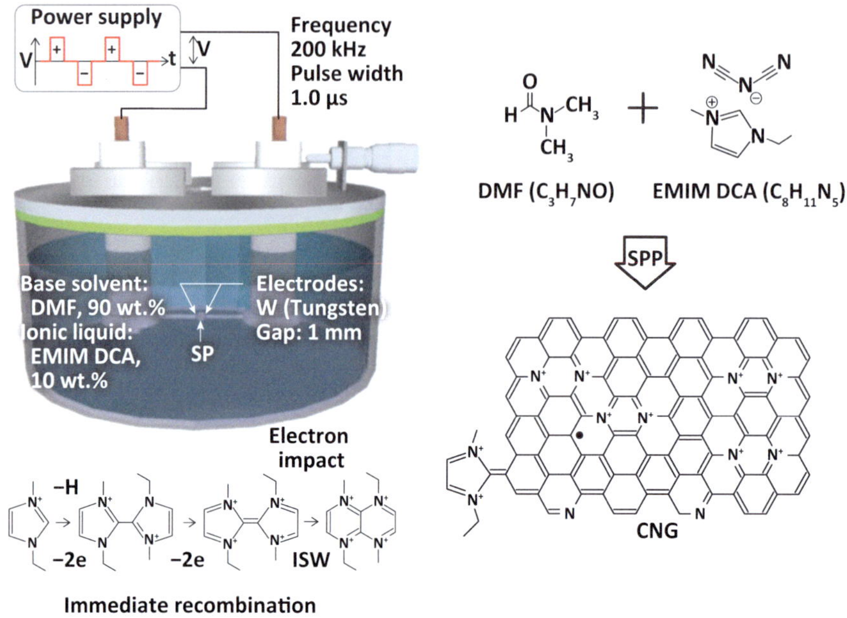

Fig. 16 Schematic image of the experimental setup for the solution plasma process (Chae et al. 2019)

many researchers are searching for Pt-free catalysts. The most promising candidates for realizing Pt-free catalysts are heteroatom-doped carbon nanomaterials.

Lee et al. (Heo et al. 2017; Lee and Saito 2018) have developed a tune-up SP system for this purpose. The unique feature of their SPP system is the fact that it employs a low-pass filter, which brings about improvement of the discharge stability. Thanks to this feature, they successfully synthesized N-doped graphitic carbon nanosheets (N-GS). Their N-GS showed a surface area as high as 621 $m^2 g^{-1}$ and nitrogen content as high as 55.5 at.% (bonded as pyridinic-N) and 21.3 at.% (bonded as pyrrolic-N). The onset potential, which is an important characteristic of ORR, was as high as 0.18 V for Ag/AgCl with a transferred electron pathway of 3.8. It should be noted that their N-GS showed very good durability, indicated by a very small current reduction of only 4% even after a long operation of 10,000 s, while Pt/C showed a very large current reduction of 22%.

Lee et al. (2017) also investigated the synthesis of the graphite doped with B and N, because such a ternary boron–carbon-nitrogen (BCN) system is also believed to be a Pt-free electrocatalyst for ORR if B and N are incorporated in the graphene network without forming B-N bonds that cause reduction of ORR activity. The actual ORR activity of the synthesized BCN nanocarbons was not as high as Pt/C. However, its durability was significantly better than Pt/C, as shown in Fig. 20, in which BCN shows a current reduction of only 15.1% after operation of 20,000 s, while Pt/C showed a very high reduction of 61.5%.

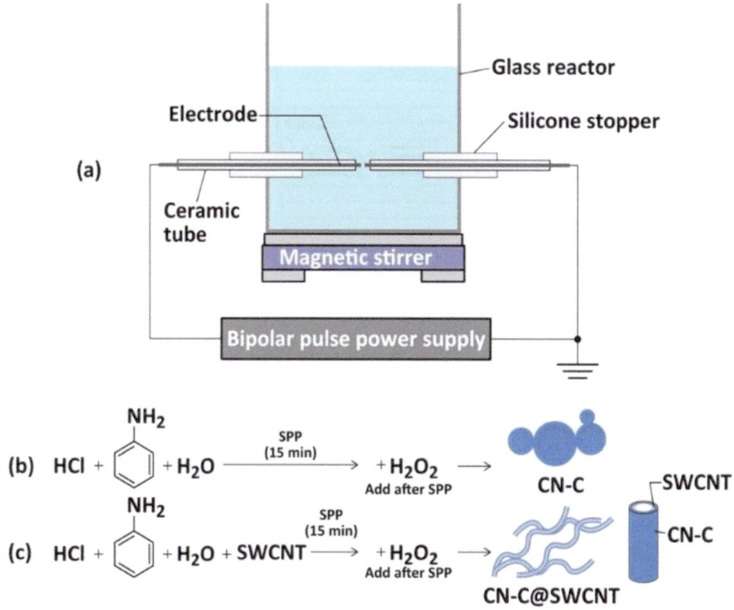

Fig. 17 **a** Experimental setup and schematics of the synthesis of **b** CN-C in an aniline aqueous solution and **c** CN-C@SWCNT in an aniline aqueous solution with the SWCNT dispersion (Chae 2019; Chae et al. 2020)

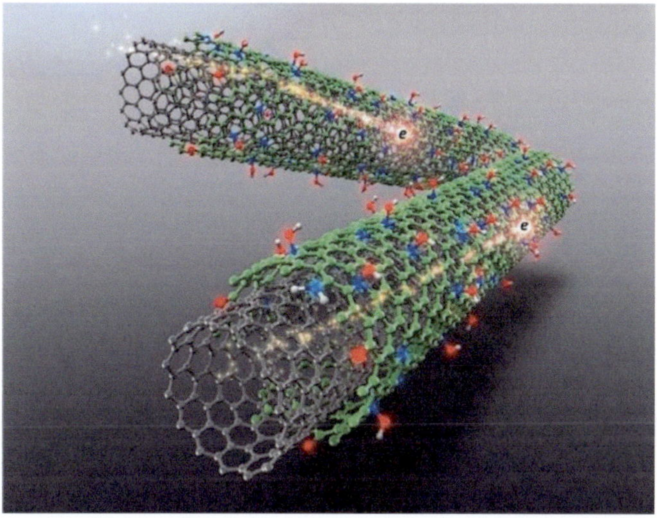

Fig. 18 Schematic of the CN-C@SWCNT in the presence of cationic nitrogen (blue) (Chae 2019)

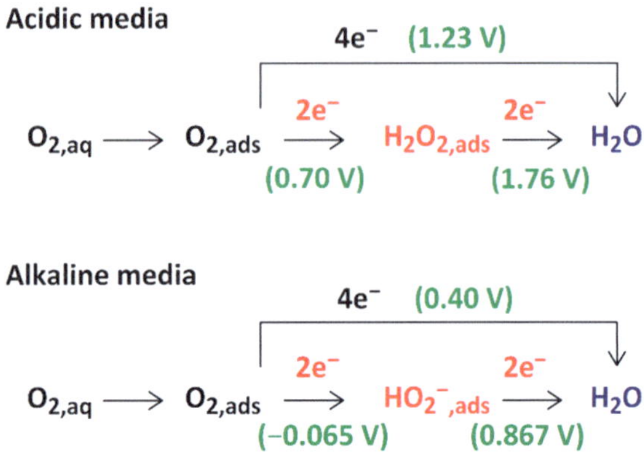

Fig. 19 The ORR pathways in the acidic and alkaline media with indication of standard electrode potential (in parentheses) (Lee 2018)

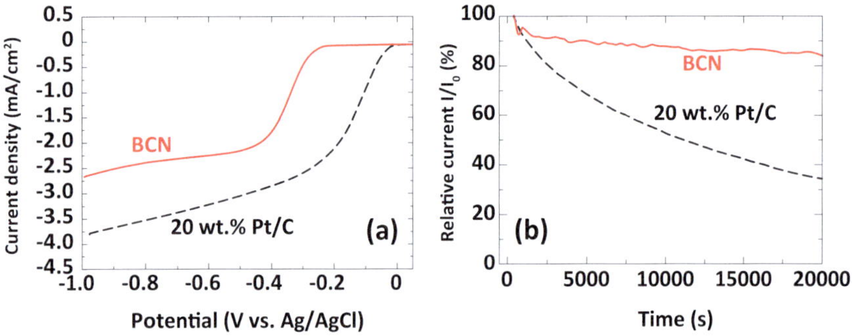

Fig. 20 **a** Linear sweep voltammetry results of BCN and 20 wt.% Pt/C in 0.1 M KOH solution at a rotation speed of 1600 rpm and a scan rate of 10 mV s^{-1}. **b** Current–time (I-t) chronoamperometry results of BCN and Pt/C at 0.4 V in O_2-saturated 0.1 M KOH solution (Lee et al. 2017; Lee 2018)

9 Mesoporous Silica

Pootawang et al. (2010, 2011a, b) proposed a unique application of SPPs. They applied the SPPs to the removal of organic surfactant templates inside mesoporous silica. The most significant feature of their method is the fact that surfactant removal can be accomplished within a very short time of 15 min in comparison to conventional calcination. In this process, active species such as OH and O generated in the SPP enter pores of mesoporous silica and decompose the templates, which are organic surfactants, in the pores, as shown in Fig. 21. Since the decomposed products are small molecules, they can be easily transported out of the pores. It should be noted that the pore structure was transformed from a disordered wormlike structure to an ordered 2D hexagonal structure at lower acidity conditions, as shown in Fig. 22. It also should be noted that the SPPs in extreme pH conditions, such as pH3 and pH11, can increase the surface area of the mesoporous silica.

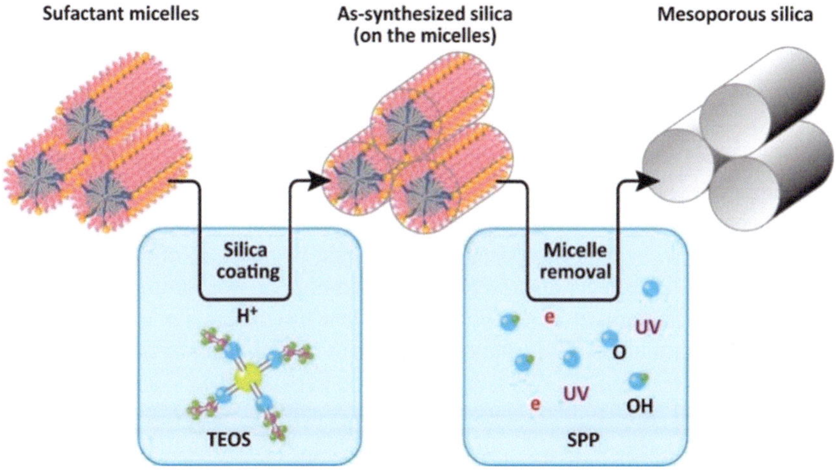

Fig. 21 Schematic representation of the synthesis of mesoporous silica by an SPP using ternary surfactant system as template and TEOS (tetraethylorthosilicate) as silica precursor (Pootawang et al. 2010; Pootawang 2011)

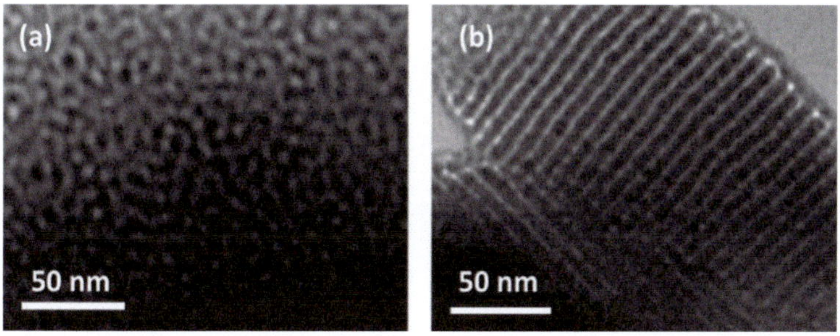

Fig. 22 **a** Disordered worm-like structures and **b** ordered 2D hexagonal structures of mesoporous silica synthesized by an SPP at pH3 solutions (Pootawang et al. 2010; Pootawang 2011)

Pootawang et al. (2012, 2012a) also found another feature of SPPs in the application of mesoporous silica production. They found that the gelation also becomes faster by using SPPs. According to their explanation, this effect is also caused by highly reactive species produced in the SPPs. In the case of conventional sol–gel methods, it takes 12 h or more. On the other hand, the SPP requires only 3 min. There is another feature regarding the shape of mesoporous silica prepared with this SPP method. In contrast to conventional sol–gel method, the SPP-prepared mesoporous silica shows a unique rope-like structure with keeping a 2D hexagonal arrangement,although detailed mechanisms are not clarified yet.

10 Low-Molecular-Weight Biopolymer

In previous section, Baroch et al. (2008) and Pootawang et al. (2010) showed that OH radicals generated by SPP play a key role in the degradation of organic compounds. Consequently, OH radicals generated by SPPs have been also applied to the depolymerization of biopolymers from high molecular weight materials to lower molecular weights for biological applications. For example, Rujiravanit et al. (2020) and Chokradjaroen et al. (2021) proposed a model for the degradation of chitin hydrogels to chitosan by OH radicals in an SPP as shown in Fig. 23. Pornsunthorntawee et al. (2014) and Prasertsung et al. (2013) reported that low molecular weight chitosan (10^3–10^4 Da) with high water solubility could be realized using an SPP.

Such depolymerization processes can also be performed by conventional thermochemical and/or enzymatic depolymerization methods. However, conventional methods are time-consuming (hours or days) and energy-consuming processes and use some toxic chemicals that are not welcome in bioapplications. The SPPs can eliminate these drawbacks associated with the conventional methods, and this is considered to be an important advantage of the SPPs.

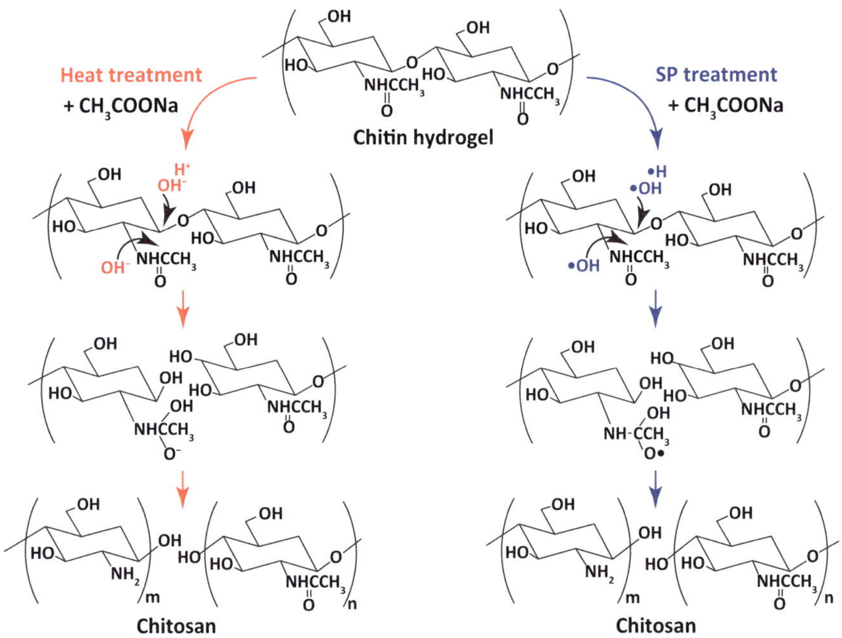

Fig. 23 The conventional thermally activated reaction schemes (left) of deacetylation and degradation of chitin hydrogel and those in the SPP (right) in an alkali condition (Rujiravanit et al. 2020; Chokradjaroen et al. 2021)

Prasertsung et al. (2012, 2013) applied the SPPs for the depolymerization of chitosan, which has attractive characteristics such as biocompatibility, biodegradability, and bioactivity. The polysaccharides such as chitosan have a chain structure in which monosaccharides are connected with glycosidic bonds. In the case of depolymerization of such macromolecules, it is important to disconnect only the glycosidic bonds without altering the unit structure of monosaccharides. According to Prasertsung et al., the SPPs can do it in the case of chitosan. The molecular weight of chitosan can be reduced from 2.5×10^5 to 6.2×10^4 Da by the SPP treatment for 30 min. They also confirmed that the SPP treatment does not cause significant fragmentation of the unit structure. It should be noted that the SPP products also involve chitooligosaccharides with a very low molecular weight of less than 10 kDa, which can be used for stem cell differentiation. Regarding the mechanism, OH and O generated in the SPP are believed to be responsible for the depolymerization reactions.

The depolymerization by the SPPs can be enhanced by adding some specific metal ions, which was revealed by Pornsunthorntawee et al. (2014). They found that the depolymerization of chitosan was significantly enhanced by adding Cu or Fe ions. However, Ag and Zn ions do not show such an effect. Enhancement of the depolymerization also occurs by using Cu or Fe electrodes instead of W electrodes, which was discovered by Tangtipapol et al. (2015). In this case, Cu or Fe ions are supplied through SP sputtering of the electrode surface. Why are only Cu and Fe ions effective in promoting depolymerization? One of the possible mechanisms is discussed in terms of the high affinity of Cu^{2+} and Fe^{3+}, which causes bond weakening when they are close enough to form coordinate bonds with chitosan. Another possible mechanism is explained in terms of Fenton reactions, in which Cu and/or Fe ions work as catalysts for producing OH radicals from H_2O_2 that is generally generated in the aqueous solution during SPPs.

Watthanaphanit and Saito (2013) applied the SPPs for the depolymerization of alginate, which is extracted from seaweed and used for various applications such as dietary supplements for improving immunity. They confirmed that the number-averaged molecular weights of alginate could be reduced from 700 to 88 kDa when the initial alginate concentration is 0.2%w/v. The depolymerization mechanism, in this case, is the same as that for chitosan. Namely, reactive oxygen species such as O and OH work to depolymerize alginates. One of the concerns regarding SPP depolymerization is its selectivity to cut only glycosidic bonds without decomposing the unit structure of monosaccharides. However, the FT-IR spectra of alginates after the SPP treatments did not show significant deviation from those before the treatment.

11 Composite Materials

A composite material is a material that combines two or more different materials expecting an improvement of various properties such as mechanical, thermal, and chemical strengths. This section describes what kinds of composites can be synthesized by using SPPs.

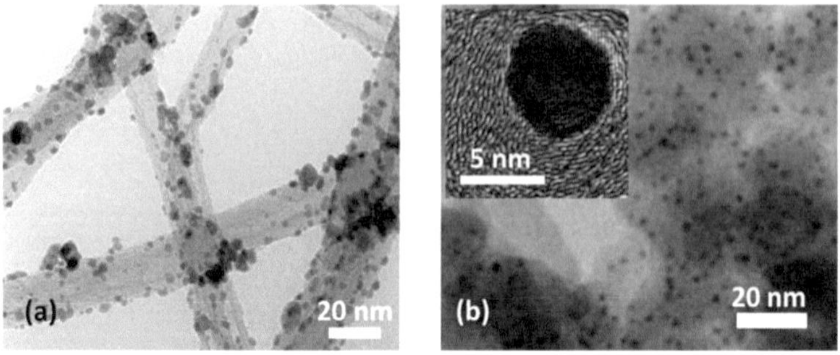

Fig. 24 TEM images of **a** Pt/MWCNT (Matsuda et al. 2014) and **b** Au/CNP (Kang et al. 2013b)

11.1 Noble Metal Nanoparticle/Carbon Composite

This type of composite has been principally investigated for electrodes in electrochemical devices such as fuel cells. Noble metal NPs for this purpose are generally Pt, Au, and Pd, and also their alloys such as PtAu, PtPd, and PtAg. These noble metal NPs can be supported by various carbon materials such as carbon nanotubes (CNTs), carbon NPs (CNP), carbon black, and graphene. There are several methods for preparing such composites, as described below.

- Mixing of separately synthesized noble metal NPs and carbon materials (Jo et al. 2016; Hu et al. 2013; Kim et al. 2014d, 2015b).
- SP synthesis of noble metal NPs in the solution containing dispersed carbon materials (Ichin et al. 2009; Yaowarat et al. 2016; Shi et al. 2014; Hu et al. 2013). A TEM image of Pt/CNT (multi-wall) composite synthesized by this method is shown in Fig. 24a (Matsuda et al. 2014).
- Combination of the bottom-up SP synthesis of noble metal NPs and the top-down SP synthesis of carbon nanomaterials using carbon electrodes (Lee et al. 2014d; Guo et al. 2009).
- Combination of the top-down SP synthesis of noble metal NPs using noble metal electrodes and the bottom-up SP synthesis of carbon nanomaterials using organic solvents such as benzene (Kang et al. 2013b; Kang and Saito 2015). It should be noted that this method has a unique feature. In the case of the other methods, NPs are found only on the surface of the carbon supports. In the case of this method, on the other hand, NPs are also embedded inside the carbon materials, as shown in Fig. 24b.

11.2 Metal Oxide/Carbon Composite

Metal oxide/carbon composites are expected to be the materials for electrochemical capacitors. The electrochemical capacitors have a superior capacity and longer life in comparison to conventional secondary batteries. Thus, they are considered as the next-generation power source for mobile devices and also electric vehicles (Chatterjee and Nandi 2021).

Lee et al. (2014c, 2015a, 2016) demonstrated the synthesis of various metal oxide/carbon composites for application to electrodes in electrochemical capacitors, also known as supercapacitors. In their method, the SPPs were used only for synthesizing metal oxide particles such as Fe_2O_3, RuO, and MnO_2. Their composites with carbon materials were prepared by depositing these metal oxide particles on carbon black.

In the case of RuO/carbon composites, their specific capacitances were higher than that of bare activated carbon and increased with an increase in the time for the SPP and oxidation. The synthesized RuO/carbon composites also showed smaller resistance and larger initial resistance slope than bare activated carbon. This effect was intensified by increasing the time spent on SPP and oxidation.

11.3 Metal Nanoparticle/Metal Oxide Composite

Noble metal NPs supported on metal oxides can be prepared through either the bottom-up and/or top-down approaches using SPPs. Such metal oxide composites coexisting with metal NPs have been studied mainly for catalysis applications because you can expect higher photocatalytic activity owing to the enhancement of charge separation and visible-light adsorption, which are caused by the localized surface plasmon resonance effect.

Panomsuwan et al. (2013) fabricated Au/$SrTiO_3$ using a top-down SPP, in which they carried out the SPP between two Au electrodes in an aqueous solution involving $SrTiO_2$ particles. Their Au/$SrTiO_3$ composites showed greater photocatalytic activity than the pure $SrTiO_3$ in the degradation of methylene blue under UV and visible light irradiation.

Mizutani et al. (2014) prepared Au/TiO_2 using an SPP in which they combined bottom-up and top-down approaches. They performed the SPP between Au electrodes in the solution of gelatinous peroxotitanic acid, in which TiO_2 was formed through bottom-up processes while Au NPs were formed through top-down processes. They revealed that most gold atoms construct the metallic nanoparticles embedded in the TiO_2 matrix from the results of characterization using UV–Vis and NEXAFS.

Nakasugi et al. (2015) prepared Au/TiO_2 using a top-down SPP, in which they carried out the SPP between Au and Pt electrodes in 0.01 M NaOH aqueous solution involving TiO_2 particles. In their experimental setup, the effective surface area of the Au electrode was designed to be much smaller than that of the Pt for generating plasma only on the Au electrode. They investigated the photocatalytic activity of the Au/TiO_2 by means of evaluating the degradation of Rhodamine B under UV irradiation for 70 min. The degradation level using the Au/TiO_2 was as high as 95%, while that of commercial TiO_2 was only 85%.

Yamamoto et al. (2015) synthesized Ag/Ga_2O_3 for CO_2 reduction with water. They conducted in situ FTIR spectroscopy to characterize the CO_2 reduction processes.

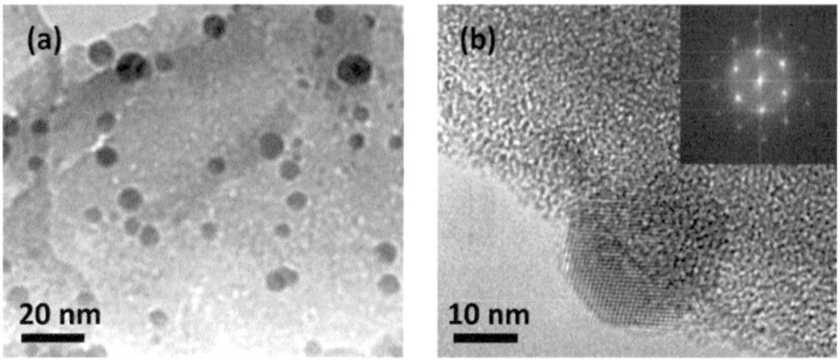

Fig. 25 **a** A TEM image of mesoporous silica with embedded AgNPs, which was synthesized by an SPP in 0.5 mM $AgNO_3$, and **b** A HRTEM image of an fcc Ag NP attached on a mesoporous silica surface (Pootawang et al. 2011a; Pootawang 2011)

According to the mechanism proposed by Yamamoto et al., the initial step of the photocatalytic reduction of CO_2 on the Ag/Ga_2O_3 is the adsorption of carbonate and bicarbonate species on the surface. These species are changed to formate species by photoirradiation. The production of CO occurs when the intermediate species interact with H_2O.

11.4 Metal Nanoparticle/Mesoporous Silica Composite

Pootawang et al. (2011a, 2012a; Pootawang and Lee 2012) synthesized the mesoporous silica with embedded Ag NPs. This material can be employed as the catalyst for reducing 4-nitrophenol or for the hydrogenation of oleic acid. To synthesize this composite material, they performed the SPP treatment of mesoporous silica in an $AgNO_3$ solution. The most important feature of this composite material is the fact that metal NPs exist not only outside mesoporous silica but also inside, which is shown in Fig. 25. This result indicates that the bottom-up reactions (namely, the reduction reaction of Ag ions) in the SPPs proceeds even in the very narrow mesopores.

11.5 Metal Nanoparticle/Biopolymer Composite

The SPPs can synthesize composites made of metal NPs and biopolymers, which can be used for biomedical applications. Synthesis of this type of composite material by using the SPPs has been demonstrated by Sang-Yul Lee's group. They investigated SP synthesis of the following biopolymers with embedded Ag NPs expecting antibacterial activity of such composites.

- gelatin (Kim et al. 2013b, 2014c; Pootawang et al. 2013a)

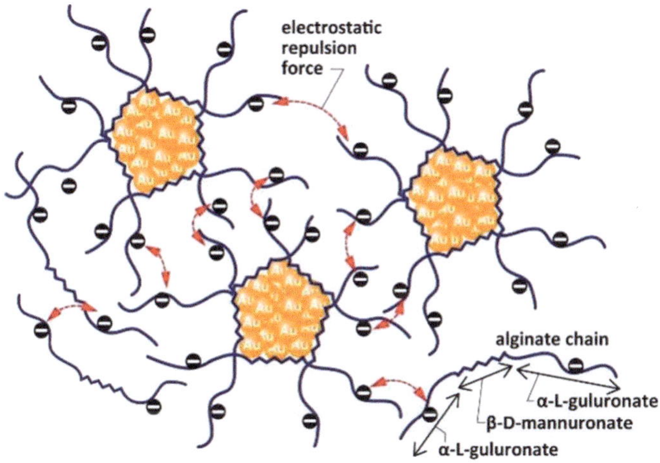

Fig. 26 Schematic illustration of Au NP suspensions produced in alginate aqueous solution (Watthanaphanit et al. 2014b)

- cellulose (Davoodbasha et al. 2015)
- chitosan (Davoodbasha et al. 2016)
- alginate (Nam et al. 2016).

They revealed that these bio-composites showed antimicrobial activity against various pathogens with minimum inhibitory concentration. Au NPs can also be embedded in biopolymers, which was demonstrated by Watthanaphanit et al. (Watthanaphanit et al. 2014a,b). They synthesized composites of alginate with embedded Au NPs for the application of cancer therapy (Fig. 26).

11.6 Polymer/Carbon Composite

Polymer/Carbon composites are also known as engineering plastics, which are fabricated by mixing polymer (as the host material) and carbon (as the filler) for the purpose of improvement in the mechanical strength of plastics with keeping their low weight. Carbon fiber-reinforced polymers, in which carbon fibers are used as the filler, have already been used practically for various applications such as automotive and airplanes.

As next-generation fillers, CNTs have attracted much attention because their mechanical strength is much stronger than that of conventional macroscopic carbon fibers. However, CNT-reinforced polymers are not yet commercially available on a large scale because there are several issues when CNTs are mixed with polymers. There are two major issues. One is an aggregation of CNTs in the host polymers. If the CNTs are not uniformly dispersed in the host material due to their aggregation,

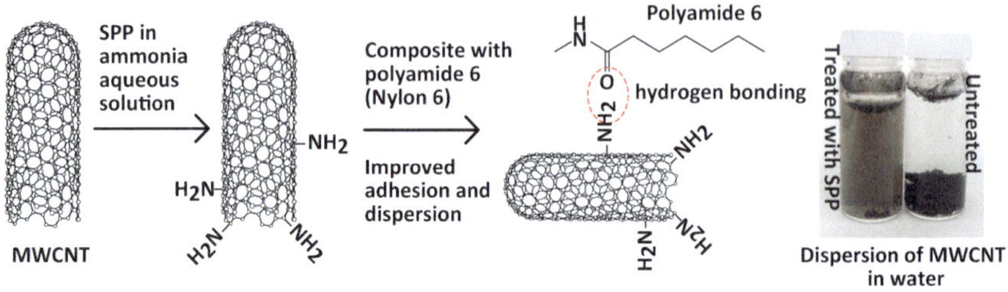

Fig. 27 Schematic illustration of preparing NH_2-functionalized CNTs (multi-wall) by the SPP under ammonia solution (Shirafuji et al. 2013). Hydrogen bonding between NH_2-functionalized CNT and oxygen in polyamide 6 brings about the improved adhesion and dispersion of the CNTs in the polyamide 6. The rightmost photograph shows the fact that the SPP-treated CNTs show better dispersion in water than as-received (untreated) CNTs

we cannot expect significant improvement in the mechanical strength of the fabricated composite material. Another issue is the hydrophobic nature of CNTs. If the CNTs are hydrophobic, we cannot expect strong bonding between the CNTs and the host material. This results in a reduction in the impact strength of the fabricated composite materials. Several attempts were made to solve these issues using SPPs.

The first successful application of SPPs in the treatment of CNTs was reported by Tong et al. (2010). The SPP treatment had two effects on the CNTs. One is to shorten the CNTs, which is thought to be caused by OH radicals generated by the SPPs. Another is to turn the CNTs to be hydrophilic, which is thought to be caused by the formation of hydrophilic groups of OH and COOH on the sidewall of the CNTs (Tong et al. 2010, 2013).

The SPP treatments of CNTs were performed not only in a acid aqueous solution but also in a base solution such as ammonia (Shirafuji et al. 2013). In this case, N-containing groups were also formed on the CNTs in addition to O-containing groups, as shown in Fig. 27. The CNTs treated with such SPPs were found to be well dispersed in water. They prepared the composite of polyamide 6 and the SPP-treated CNTs. The SPP-treated CNTs were also found to be well dispersed in polyamide 6, while as-received CNTs were aggregated in polyamide 6. Thanks to these improved properties of the SPP-treated CNTs, the mechanical strength of the composite prepared with the SPP-treated CNTs was better than those with as-received CNTs.

12 Metal–Organic Framework

Metal complexes have been applied as functional materials in a wide range of fields, such as medical science and chemical industries. In particular, metal–organic frameworks (MOFs) are porous metal complex materials that are expected to be applied to catalysts and sensors due to their high surface area and designability. Various

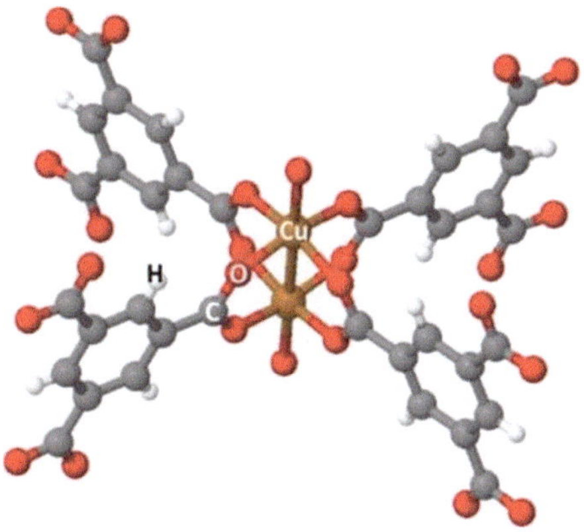

Fig. 28 Molecular structure of HKUST-1

methods have been developed for their synthesis, including solvothermal and electrochemical synthesis, and recently, a synthesis method using plasma/discharge has been reported (Tao et al. 2019).

Kanno et al. (2021a, b) have verified the effectiveness of the properties of plasma/discharge, such as reactivity and charging, on the synthesis and deposition processes of metal complexes. They conducted various analyses of the synthesized products to observe the effect of reactivity in the synthesis of metal complexes by SPPs; plasma was generated in a liquid in which metal ions and organic ligands were dissolved. What they synthesized was HKUST-1 (Fig. 28), which is assembled from Cu(II) nodes and BTC (1,3,5-benzenetricarboxylate). In their system, the SPP was performed by using a bipolar pulse power source (80 kHz, 1.5 kV) on the electrode. The solution used in their study was an ethanol–water (30 ml and 10 ml) solution containing $Cu(NO_3)_2$ ($3H_2O$) (0.34 g) and H3BTC (0.2 g). They compared HKUST-1 synthesized by their SPP to those by the conventional methods; one is a heating method (also known as a solvothermal method), and another is the addition of triethylamine.

From XRD and FT-IR spectroscopy, they confirmed that the product produced by the SPP was HKUST-1. Furthermore, the HKUST-1 synthesized by the SPP was found to contain monovalent copper ions from fluorescence spectroscopy and XPS, which indicated that the SPP could synthesize mixed-valence MOFs differently from other conventional methods. The HKUST-1 by the SPP has another significant feature in the water resistance characteristics. They immersed three types of HKUST-1, which were synthesized by the SPP, heating, and triethylamine, in water for 12 h. The water tolerance was characterized by XRD, of which results are shown in Fig. 29. Two HKUST-1 synthesized by the conventional methods showed structural changes, while HKUST-1 synthesized by the SPP retained its crystallinity even after immersion in water. The reason for this improvement in water tolerance may be that the plasma in the liquid makes HKUST-1 contain mixed-valence metal ions and that the adsorption of water molecules is inhibited by the adsorption of plasma-derived active species on the open metal sites of HKUST-1 (Todaro et al. 2016). The effect of the plasma

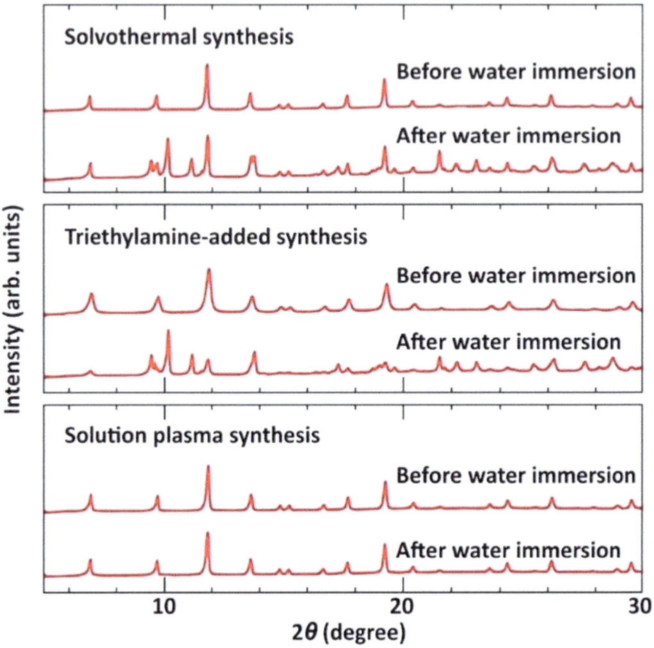

Fig. 29 Chemical stability of MOFs synthesized by solvothermal method, the addition of triethylamine, and solution plasma (Kanno et al. 2021b; Kanno 2022)

on the organic ligands could also be one of the factors for the improvement of water tolerance.

13 Summary

Solution plasma is a non-thermal equilibrium plasma in solution and is a low-temperature plasma reaction field, so its application to various material synthesis is being promoted. In the material synthesis process using solution plasma, new reactions and new material synthesis that cannot be realized by conventional solution reactions are possible by controlling the active species to be generated and the reaction space. In particular, remarkable research results have been produced on the synthesis of nanoparticles, carbon surface modification, and the synthesis of carbon materials and heterocarbon materials. It is expected that a new field of chemistry in solution will be opened up by realizing a more precise material synthesis process.

References

Ashkarran AA (2010) A novel method for synthesis of colloidal silver nanoparticles by arc discharge in liquid. Curr Appl Phys 10:1442–1447

Ashkarran A (2011) Metal and metal oxide nanostructures prepared by electrical arc discharge method in liquids. J Cluster Sci 22:233–266

Ashkarran AA (2012) Synthesis and characterization of gold nanoparticles via submerged arc discharge based on a seed-mediated approach. J Theor Appl Phys 6(14):1–6

Ashkarran AA, Zad AI, Ahadian MM, Ardakani SAM (2008) Synthesis and photocatalytic activity of WO_3 nanoparticles prepared by the arc discharge method in deionized water. Nanotechnol 19(195709):1–7

Ashkarran AA, Zad AI, Mahdavi SM, Ahadian MM, Nezhad MRH (2009a) Rapid and efficient synthesis of colloidal gold nanoparticles by arc discharge method. Appl Phys A 96:423–428

Ashkarran AA, Zad AI, Mahdavia SM, Ahadian MM (2009b) ZnO nanoparticles prepared by electrical arc discharge method in water. Mater Chem Phys 118:6–8

Ashkarran A, Kavianipour M, Aghigh S, Afshar SA, Saviz S, Zad AI (2010a) On the formation of TiO_2 nanoparticles via submerged arc discharge technique: synthesis, characterization and photocatalytic properties. J Cluster Sci 21:753–766

Ashkarran AA, Zad AI, Mahdavi SM, Ahadian MM (2010c) Photocatalytic activity of ZnO nanoparticles prepared via submerged arc discharge method. Appl Phys A 100:1097–1102

Ashkarran AA, Afshar SAA, Aghigh SM, kavianipour M (2010b) Photocatalytic activity of ZrO_2 nanoparticles prepared by electrical arc discharge method in water. Polyhedron 29:1370–1374

Baroch P, Anita V, Saito N, Takai O (2008) Bipolar pulsed electrical discharge for decomposition of organic compounds in water. J Electrostat 66:294–299

Bera D, Kuiry SC, McCutchen M, Seal S, Heinrich H, Slane GC (2004) In situ synthesis of carbon nanotubes decorated with palladium nanoparticles using arc-discharge in solution method. J Appl Phys 96:5152–5157

Bohnart ER (2018) Welding principles and practices, 5th edn. McGraw-Hill, New York, NY

Bratescu MA, Saito N (2013) Charge doping of large-area graphene by gold-alloy nanoparticles. J Phys Chem C 117:26804–26810

Bratescu M, Saito N, Takai O (2011a) Redox reactions in liquid plasma during iron oxide and oxide-hydroxide nanoparticles synthesis. Curr Appl Phys 11:S30–S34

Bratescu MA, Cho SP, Takai O, Saito N (2011b) Size-controlled gold nanoparticles synthesis in solution plasma. J Phys Chem C 115:24569–24576

Bratescu MA, Takai O, Saito N (2013) One-step synthesis of gold bi metallic nanoparticles with various m etal-compositions. J Alloys Compd 562:74–83

Brust M, Walker M, Bethell D, Schiffrin DJ, Whyman R (1994) Synthesis of thiol-derivatised gold nanoparticles in a two-phase liquid-liquid system. J Chem Soc, Chem Commun 801–802

Burakov VS, Nevar AA, Nedelko MI, Tarasenko VN (2008) Formation of zinc oxide nanoparticles during electric discharge in water. Tech Phys Lett 34:679–681

Chae S, Bratescu MA, Saito N (2017) Synthesis of few-layer graphene by peeling graphite flakes via electron exchange in solution plasma. J Phys Chem C 121:23793–32302

Chae S, Panomsuwan G, Bratescu MA, Teshima K, Saito N (2019) P-type doping of graphene with cationic nitrogen. ACS Appl Nano Mater 2:1350–1355

Chae S, Phan PQ, Panomsuwan G, Bratescu MA, Hashimoto T, Teshima K, Saito N (2020) Single-walled carbon nanotubes wrapped by cationic nitrogen-doped carbon for electrocatalytic applications. ACS Appl Nano Mater 3:10183–10189

Chae S (2019) Synthesis and properties of cationic nitrogen-doped carbon materials via solution plasma. PhD thesis, Nagoya University, Nagoya, Japan

Chai GL, Hou Z, Shu DJ, Ikeda T, Terakura K (2014) Active sites and mechanisms for oxygen reduction reaction on nitrogen-doped carbon alloy catalysts: stone-wales defect and curvature effect. J Am Chem Soc 136(39):13629–13640

Chatterjee DP, Nandi AK (2021) A review on the recent advances in hybrid supercapacitors. J Mater Chem A 9:15880–15918

Chen L, Mashimo T, Omurzak E, Okudera H, Iwamoto C, Yoshiasa A (2011) Pure tetragonal ZrO_2 nanoparticles synthesized by pulsed plasma in liquid. J Phys Chem C 115:9370–9375

Chen Q, Kaneko T, Hatakeyama R (2012) Synthesis of superfine ethanol-soluble CoO nanoparticles via discharge plasma in liquid. Appl Phys Express 5(096201):1–3

Cho C, Choi Y, Kang C, Lee G (2007) Effects of the medium on synthesis of nanopowders by wire explosion process. App Phy Lett 91(141501):1–3

Cho SP, Bratescu MA, Saito N, Takai O (2011) Microstructural characterization of gold nanoparticles synthesized by solution plasma processing. Nanotechnol 22(455701):1–7

Chokradjaroen C, Niu J, Panomsuwan G, Saito N (2021) Insight on solution plasma in aqueous solution and their application in modification of chitin and chitosan. Int J Mol Sci 22(4308):1–23

Davoodbasha M, Lee SY, Kim SC, Kim JW (2015) One-step synthesis of cellulose/silver nanobio-composites using a solution plasma process and characterization of their broad spectrum antimicrobial efficacy. RSC Adv 5:35052–35060

Davoodbasha M, Kim SC, Lee SY, Kim JW (2016) The facile synthesis of chitosan-based silver nano-biocomposites via a solution plasma process and their potential antimicrobial efficacy. Arch Biochem Biophys 605:49–58

Fisher KB, Thagard SM (2012) Transformation of organic solvents into carbon-based materials by liquid-phase plasmas. Plasma Chem Plasma Process 32:919–931

Fridman A (2008) Plasma chemistry. Cambridge University Press, Cambridge

Fu SY, Li YZ, Chu W, Yang YM, Tong DG, Zeng QL (2015) Retracted article: mesoporous amorphous FeOF nanococoons for high-rate and long-life rechargeable sodium-ion batteries. J Mater Chem A 3:16716–16727

Guo J, Wang X, Xu B (2009) One-step synthesis of carbon-onion-supported platinum nanoparticles by arc discharge in an aqueous solution. Mater Chem Phys 113:179–182

Guo HL, Su P, Kang X, Ning SK (2013) Synthesis and characterization of nitrogen-doped graphene hydrogels by hydrothermal route with urea as reducing-doping agents. J Mater Chem A 1:2248–2255

Hagino T, Kondo H, Ishikawa K, Kano H, Sekine M, Hori M (2012) Ultrahigh-speed synthesis of nanographene using alcohol in-liquid plasma. Appl Phys Express 5(035101):1–3

He YJ, Peng JF, Chu W, Li YZ, Tong DG (2014) Retracted article: black mesoporous anatase TiO_2 nanoleaves: a high capacity and high rate anode for aqueous Al-ion batteries. J Mater Chem A 2:1721–1731

Heberlein J, Murphy AB (2008) Thermal plasma waste treatment. J Phys D 41(053001):1–20

Heo YK, Lee SY (2011) Effects of the gap distance on the characteristics of gold nanoparticles in nanofluids synthesized using solution plasma processing. Met Mater Int 17:431–434

Heo YK, Bratescu MA, Ueno T, Saito N (2014) Synthesis of mono-dispersed nanofluids using solution plasma. J Appl Phys 116(024302):1–7

Heo YK, Lee SH, Bratescu MA, Kim SM, Lee GJ, Saito N (2017) Generation of non-equilibrium condition in solution plasma discharge using low-pass filter circuit. Plasma Process Polym 14(e1600163):1–7

Hieda J, Saito N, Takai O (2008a) Exotic shapes of gold nanoparticles synthesized using plasma in aqueous solution. J Vac Sci Technol A 26:854–856

Hieda J, Saito N, Takai O (2008b) Size-regulated gold nanoparticles fabricated by a discharge in reverse micelle solutions. Surf Coat Technol 202:5343–5346

Hu X, Cho SP, Takai O, Saito N (2012) Rapid synthesis and structural characterization of well-defined gold clusters by solution plasma sputtering. Cryst Growth des 12:119–123

Hu X, Shen X, Takai O, Saito N (2013) Facile fabrication of PtAu alloy clusters using solution plasma sputtering and their electrocatalytic activity. J Alloys Compd 552:351–355

Hu X, Zhang X, Shen X, Li H, Takai O, Saito N (2014) Plasma-induced synthesis of CuO nanofibers and zno nanoflowers in water. Plasma Chem Plasma Process 34:1129–1139

Hu X, Shi J, Zhang J, Tang W, Zhu H, Shen X, Saito N (2015) One-step facile synthesis of carbon-supported PdAu nanoparticles and their electrochemical property and stability. J Alloys Compd 619:452–457

Hyun K, Ueno T, Li OL, Saito N (2016a) Synthesis of heteroatom-carbon nanosheets by solution plasma processing using N-methyl-2-pyrrolidone as precursor. RSC Adv 6:6690–6696

Hyun K, Ueno T, Panomsuwan G, Li OL, Saito N (2016b) Heterocarbon nanosheets incorporating iron phthalocyanine for oxygen reduction reaction in both alkaline and acidic media. Phys Chem Chem Phys 18:10856–10863

Hyun K, Ueno T, Saito N (2016c) Synthesis of nitrogen-containing carbon by solution plasma in aniline with high-repetition frequency discharges. Jpn J Appl Phys 55:01AE18(1–6pp)

Ichin Y, Mitamura K, Saito N, Takai O (2009) Characterization of platinum catalyst supported on carbon nanoballs prepared by solution plasma processing. J Vac Sci Technol A 27:826–830

Ishizaki T, Chiba S, Kaneko Y, Panomsuwan G (2014) Electrocatalytic activity for the oxygen reduction reaction of oxygen-containing nanocarbon synthesized by solution plasma. J Mater Chem A 2:10589–10598

ISO (2015) International Organization for Standardization: Nanotechnologies—Vocabulary—Part 1: Core terms. (ISO Standard no. 80004-1:2015). https://www.iso.org/obp/ui/#iso:std:iso:ts:80004:-1:ed-2:v1:en

Jameson EC (2001) Electrical discharge machining. Society of Manufacturing Engineers, Dearbron, MI

Jedsukontorn T, Ueno T, Saito N, Hunsom M (2017) Facile preparation of defective black TiO_2 through the solution plasma process: effect of parametric changes for plasma discharge on its structural and optical properties. J Alloys Compd 726:567–577

Jedsukontorn T, Ueno T, Saito N, Hunsom M (2018) Narrowing band gap energy of defective black TiO_2 fabricated by solution plasma process and its photocatalytic activity on glycerol transformation. J Alloys Compd 757:188–199

Jo Y G, Kim S M, Lee S Y (2016) Size-tunable palladium nanoparticles synthesized using the solution plasma process and their electrocatalytic activities. Jpn J Appl Phys 55:01AE01(1–5pp)

Kang J, Saito N (2015) In situ solution plasma synthesis of mesoporous nanocarbon-supported bimmetallic nanoparticles. RSC Adv 5:29131–29134

Kang J, Li OL, Saito N (2013a) Hierarchical meso-macro structure porous carbon black as electrode materials in Li-air battery. J Power Sources 261:156–161

Kang J, Li OL, Saito N (2013b) A simple synthesis method for nano-metal catalyst supported on mesoporous carbon: the solution plasma process. Nanoscale 5:6874–6882

Kang J, Li OL, Saito N (2013c) Synthesis of structure-controlled carbon nano spheres by solution plasma process. Carbon 60:292–298

Kanno M, Ito T, Shimizu Y, Terashima K (2021a) Cuboid $Cu(HBTC)(H_2O)_3$ synthesis via plasma pretreatment of trimesic acid solution. Plasma Process Polym 18:2100047

Kanno M, Kitao T, Ito T, Terashima K (2021b) Synthesis of a metal-organic framework by plasma in liquid to increase reduced metal ions and enhance water stability. RSC Adv 11:22756–22760

Kanno M (2022) Synthesis and deposition of metal complexes with plasma/discharge in liquid. PhD thesis, The University of Tokyo, Tokyo, Japan

Kelgenbaeva Z, Omurzak E, Takebe S, Sulaimankulova S, Abdullaeva Z, Iwamoto C, Mashimo T (2014) Synthesis of pure iron nanoparticles at liquid-liquid interface using pulsed plasma. J Nanopart Res 16(2603):1–11

Kelgenbaeva Z, Omurzak E, Ihara H, Iwamoto C, Sulaimankulova S, Mashimo T (2015) Sn and SnO_2 nanoparticles by pulsed plasma in liquid: synthesis, characterization and applications. Phys Status Solidi A 212:2951–2957

Kelgenbaeva Z, Omurzak E, Takebe S, Abdullaeva Z, Sulaimankulova S, Iwamoto C, Mashimo T (2013) Magnetite nanoparticles synthesized using pulsed plasma in liquid. Jpn J Appl Phys 52:11NJ02(5pp)

Kibria G, Jahan M P, Bhattacharyya B (eds) (2019) In: Micro-electrical discharge machining processes. Springer Nature, Singapore

Kim SM, Kim GS, Lee SY (2008) Effects of PVP and KCl concentrations on the synthesis of gold nanoparticles using a solution plasma processing. Mater Lett 62:4354–4356
Kim SC, Kim JW, Yoon GJ, Nam SW, Lee SY (2013b) Antifungal effects of 3D scaffold type gelatin/Ag nanoparticles biocomposite prepared by solution plasma processing. Curr Appl Phys 13:S48–S53
Kim DW, Li OL, Pootawang P, Saito N (2014a) Solution plasma synthesis process of tungsten carbide on N-doped carbon nanocomposite with enhanced catalytic ORR activity and durability. RSC Adv 4:16813–16819
Kim DW, Li OL, Saito N (2014b) The role of the central Fe atom in the N4-macrocyclic structure for the enhancement of oxygen reduction reaction in a heteroatom nitrogen-carbon nanosphere. Phys Chem Chem Phys 16:14905–14911
Kim SC, Kim SM, Yoon GJ, Nam SW, Lee SY, Kim JW (2014c) Gelatin-based sponge with Ag nanoparticles prepared by solution plasma: Fabrication, characteristics, and their bactericidal effect. Curr Appl Phys 14:S172–S179
Kim SM, Lee YJ, Kim JW, Lee SY (2014d) Facile synthesis of Pt-Pd bim etallic nanoparticles by plasma discharge in liquid and their electrocatalytic activity toward methanol oxidation in alkaline media. Thin Solid Films 572:260–265
Kim DW, Li OL, Saito N (2015a) Enhancement of ORR catalytic activity by multiple heteroatom-doped carbon materials. Phys Chem Chem Phys 17:407–413
Kim SM, Jo YG, Lee SY (2015b) The composition-controlled synthesis of Pt-Ag bim etallic nanochains for catalytic methanol oxidation. Electrochim Acta 174:1244–1252
Kim H, Watthanaphanit A, Saito N (2016) Synthesis of colloidal MnO_2 with a sheet-like structure by one-pot plasma discharge in permanganate aqueous solution. RSC Adv 6:2826–2834
Kim K, Chae S, Choi PG, Itoh T, Saito N, Masuda Y (2021) Facile synthesis of ZnO nanobullets by solution plasma without chemical additives. RSC Adv 11:26785–26790
Kim HG, Lee H, Kim BH, Kim SJ, Lee JM, Jung SC (2013a) Synthesis process of cobalt nanoparticles in liquid-phase plasma. Jpn J Appl Phys 52(01AN03):1–4
Koo IG, Lee MS, Shim JH, Ahn JH, Lee WM (2005) Platinum nanoparticles prepared by a plasma-chemical reduction method. J Mater Chem 15:4125–4128
Kroushawi F, Latifi H, Hosseini SH, Peysokhan M, Nikbakht H, Silani Y, Ghomi H (2012) Characterization of RF discharge in liquid n-hexane and its application to synthesize carbon nano-particles. Plasma Chem Plasma Process 32:959–968
Lal A, Bleuler H, Wüthrich R (2008) Fabrication of metallic nanoparticles by electrochemical discharges. Electrochem Commun 10:488–491
Lange H, Sioda M, Huczko A, Zhu Y, Kroto H, Walton D (2003) Nanocarbon production by arc discharge in water. Carbon 41:1617–1623, cited by 183
Lee SH, Saito N (2018) Enhancement of nitrogen self-doped nanocarbons electrocatalyst via tune-up solution plasma synthesis. RSC Adv 8:35503–35511
Lee H, Park SH, Jung SC, Yun JJ, Kim SJ, Kim DH (2013) Preparation of nonaggregated silver nanoparticles by the liquid phase plasma reduction method. J Mater Res 28:1105–1110
Lee H, Bratescu MA, Ueno T, Saito N (2014a) Solution plasma exfoliation of graphene flakes from graphite electrodes. RSC Adv 4:51758–51765
Lee H, Park S, Kim S, Park Y, Kim B, Jung S (2014b) Synthesis of tin and tin oxide nanoparticles using liquid phase plasma in an aqueous solution. Microelectron Eng 126:153–157
Lee YJ, Kim SM, Kim JW, Lee SY (2014d) The characterization and electrocatalytic activities of carbon-supported Pt nanoparticles synthesized by the solution plasma process. Mater Lett 123:184–186
Lee H, Park SH, Kim SJ, Park YK, Kim BJ, An KH, Ki SJ, Jung SC (2015a) Synthesis of manganese oxide/activated carbon composites for supercapacitor application using a liquid phase plasma reduction system. Int J Hydrogen Energy 40:754–759
Lee H, Ueno T, Saito N (2015b) The effect of electrode gap distance on the synthesis of carbon materials by using solution plasma process. J Miner Met Mater Soc 67:2550–2556

Lee H, Kim SJ, An KH, Kim JS, Kim BH, Jung SC (2016) Application of liquid phase plasma process to the synthesize ruthenium oxide/activated carbon composite as dielectric material for supercapacitor. Adv Mater Lett 7:98–103
Lee SH, Heo YK, Bratescu MA, Uenoa T, Saito N (2017) Solution plasma synthesis of a boron-carbon-nitrogen catalyst with a controllable bond structure. Phys Chem Chem Phys 19:15264–15272
Lee H, Park SH, Kim SJ, Park YK, An KH, Kim BJ, Jung SC (2014c) Liquid phase plasma synthesis of iron oxide/carbon composite as dielectric material for capacitor. J Nanomater 2014:132032(1–6pp)
Lee H (2016) Control of nano carbon structure by using solution plasma process. PhD thesis, Nagoya University, Nagoya, Japan
Lee S (2018) Solution plasma synthesis of heterographene for energy conversion system. PhD thesis, Nagoya University, Nagoya, Japan
Li HOL, Kang J, Urashima K, Saito N (2013a) Comparison between the mechanism of liquid plasma discharge process in water and organic solution. J Inst Electrostat Jpn 37:22–27
Li O, Kang J, Urashima K, Saito N (2013b) Solution plasma synthesis process of carbon nano particles in organic solutions. Int J Plasma Environ Sci Technol 7:31–36
Li OL, Hayashi H, Ishizaki T, Saito N (2016) Enhancement of conductivity in nano carbon balls by the addition of carbon tetrachloride via room temperature solution plasma process. RSC Adv 6:51864–51870
Li OL, Shi Z, Lee H, Ishizaki T (2019) Enhanced electrocatalytic stability of platinum nanoparticles supported on sulfur-doped carbon using in-situ solution plasma. Sci Rep 9(12704):1–10
Li O L, Lee H, Ishizaki T (2018) Recent progress in solution plasma-synthesized-carbon-supported catalysts for energy conversion systems. Jpn J Appl Phys 5:0102A2 (1–14pp)
Lieberman MA, Lichtenberg AJ (2005) Principles of plasma discharges and materials processing, 2nd edn. Wiley-Interscience, Hoboken, NJ
Lipshutz BH, Ghorai S (2008) Transition-metal-catalyzed cross-couplings going green: in water at room temperature. Aldrichimica Acta 41:59–72
Lo CH, Tsung TT, Chen LC (2005) Shape-controlled synthesis of Cu-based nanofluid using submerged arc nanoparticle synthesis system (SANSS). J Cryst Growth 277:636–642
Matsuda N, Nakashima T, Kato T, Shiroishi H (2014) Synthesis of multiwall carbon nanotube-supported platinum catalysts by solution plasma processing for oxygen reduction in polymer electrolyte fuel cells. Electrochim Acta 146:73–78
Matsushima M, Noda M, Yoshida T, Kato H, Kalita G, Kizuki T, Uchida H, Umeno M, Wakita K (2013) Formation of graphene nano-particle by means of pulsed discharge to ethanol. J Appl Phys 113:114304(4pp)
Mingos DMP (2014a) Gold clusters, colloids and nanoparticles I. Springer, Switzerland, pp 1–48
Mingos DMP (2014b) In: Gold clusters, colloids and nanoparticles II. Springer, Switzerland, pp 1–66
Miron C, Bratescu MA, Saito N, Takai O (2010) Time-resolved optical emission spectroscopy in water electrical discharges. Plasma Chem Plasma Process 30:619–631
Miron C, Bratescu MA, Saito N, Takai O (2011) Optical diagnostic of bipolar electrical discharges in HCl, KCl, KOH solutions. J Appl Phys 109(123301):1–8
Mizutani T, Murai T, Nomoto T, Nameki H, Yoshida T, Yagi S (2014) Preparation of gold nanoparticles included in anatase-phase titania by solution plasma method. Surf Interface Anal 46:1125–1128
Morishita T, Ueno T, Panomsuwan G, Hieda J, Yoshida A, Bratescu MA, Saito N (2016) Fastest formation routes of nanocarbons in solution plasma processes. Sci Rep 6(36880):1–13
Nakasugi Y, Saito G, Yamashita T, Akiyama T (2014) Synthesis of nonstoichiometric titanium oxide nanoparticles using discharge in HCl solution. J Appl Phys 115(123303):1–7
Nakasugi Y, Saito G, Yamashita T, Sakaguchi N, Akiyama T (2015) Solution plasma synthesis of Au nanoparticles for coating titanium dioxide to enhance its photocatalytic activity. Thin Solid Films 583:135–141

Nam S, MubarakAli D, Kim J (2016) Characterization of alginate/silver nanobiocomposites synthesized by solution plasma process and their antimicrobial properties. J Nanomater 4712813:1–9
Omurzak E, Abdullaeva Z, Iwamoto C, Ihara H, Sulaimankulova S, Mashimo T (2015) Synthesis of hollow carbon nano-onions using the pulsed plasma in liquid. J Nanosci Nanotechnol 15:3703–3709
Panomsuwan G, Saito N, Ishizaki T (2006) Nitrogen-doped carbon nanoparticle-carbon nanofiber composite as an efficient metal -free cathode catalyst for oxygen reduction reaction. ACS Appl Mater Interfaces 8:6962–6971
Panomsuwan G, Chiba S, Kaneko Y, Saito N, Ishizaki T (2014) In situ solution plasma synthesis of nitrogen-doped carbon nanoparticles as metal-free electrocatalysts for the oxygen reduction reaction. J Mater Chem A 2:18677–18686
Panomsuwan G, Saito N, Ishizaki T (2015a) Electrocatalytic oxygen reduction activity of boron-doped carbon nanoparticles synthesized via solution plasma process. Electrochem Commun 59:81–85
Panomsuwan G, Saito N, Ishizaki T (2015b) Nitrogen-doped carbon nanoparticles derived from acrylonitrile plasma for electrochemical oxygen reduction. Phys Chem Chem Phys 17:6227–6323
Panomsuwan G, Saito N, Ishizaki T (2015c) Simple one-step synthesis of fluorine-doped carbon nanoparticles as potential alternative metal-free electrocatalysts for oxygen reduction reaction. J Mater Chem A 3:9972–9981
Panomsuwan G, Watthanaphanit A, Ishizaki T, Saito N (2015d) Water-plasma-assisted synthesis of black titania spheres with efficient visible-light photocatalytic activity. Phys Chem Chem Phys 17:13794–13799
Panomsuwan G, Saito N, Ishizaki T (2016) Electrocatalytic oxygen reduction on nitrogen-doped carbon nanoparticles derived from cyano-aromatic molecules via a solution plasma approach. Carbon 98:411–420
Panomsuwan G, Zettsu N, Saito N (2013) Gold nanoparticles supported on $SrTiO_3$ by solution plasma sputter deposition for enhancing UV- and visible-light photocatalytic efficiency. Mater Res Soc Symp Proc 1509:mrsf12–1509–cc09–14
Panomsuwana G, Saito N, Ishizaki T (2016) From cyano-aromatic molecules to nitrogen-doped carbons by solution plasma for the oxygen reduction reaction in alkaline medium. Mater Today Proc 2:4302–4308
Pootawang P, Lee SY (2012) Rapid synthesis of Ag nanoparticles-embedded mesoporous silica via solution plasma and its catalysis for 4-nitrophenol reduction. Mater Lett 80:1–4
Pootawang P, Saito N, Takai O (2010) Solution plasma process for template removal in mesoporous silica synthesis. Jpn J Appl Phys 49(126202):1–7
Pootawang P, Saito N, Takai O (2011a) Ag nanoparticle incorporation in mesoporous silica synthesized by solution plasma and their catalysis for oleic acid hydrogenation. Mater Lett 65:1037–1040
Pootawang P, Saito N, Takai O (2011b) Solution plasma for template removal in mesoporous silica: pH and discharge time varying characteristics. Thin Solid Films 519:7030–7035
Pootawang P, Saito N, Takai O, Lee SY (2012a) Rapid synthesis of ordered hexagonal mesoporous silica and their incorporation with Ag nanoparticles by solution plasma. Mater Res Bull 47:2726–2729
Pootawang P, Saito N, Takai O, Lee SY (2012b) Synthesis and characteristics of Ag/Pt bi metallic nanocomposites by arc-discharge solution plasma processing. Nanotechnol 23(395602):1–8
Pootawang P, Kim SC, Kim JW, Lee SY (2013a) Incorporation of Ag metal lic nanoparticles in 3D gelatin matrix via the green strategy solution plasma. J Nanosci Nanotechnol 13:589–592
Pootawang P, Saito N, Lee SY (2013b) Discharge time dependence of a solution plasma process for colloidal copper nanoparticle synthesis and particle characteristics. Nanotechnol 24(055604):1–9

Pootawang P (2011) Solution plasma synthesis and characterization for mesoporous silica and metal nanoparticles system. PhD thesis, Nagoya University, Nagoya, Japan

Pornsunthorntawee O, Katepetch C, Vanichvattanadecha C, Saito N, Rujiravanit R (2014) Depolymerization of chitosan-metal complexes via a solution plasma technique. Carbohydr Polym 102:504–512

Potoký Š, Saito N, Takai O (2009) Needle electrode erosion in water plasma discharge. Thin Solid Films 518:918–923

Prasertsung I, Damrongsakkul S, Terashima C, Saito N, Takai O (2012) Preparation of low molecular weight chitosan using solution plasma system. Carbohydr Polym 87:2745–2749

Prasertsung I, Damrongsakkul S, Saito N (2013) Degradation of beta-chitosan by solution plasma process (SPP). Polym Degrad Stab 98:2089–2093

Rujiravanit R, Kantakanun M, Chokradjaroen C, Vanichvattanadecha C, Saito N (2020) Simultaneous deacetylation and degradation of chitin hydrogel by electrical discharge plasma using low sodium hydroxide concentrations. Carbohydr Polym 228(115377):1–9

Safa S, Soucy G (2014) Liquid and solution treatment by thermal plasma: a review. Int J Environ Sci Technol 11:1165–1188

Saito N, Hieda J, Takai O (2009) Synthesis process of gold nanoparticles in solution plasma. Thin Solid Films 518:912–917

Saito G, Hosokai S, Akiyama T, Yoshida S, Yatsu S, Watanabe S (2010) Size-controlled Ni nanoparticles formation by solution glow discharge. J Phys Soc Jpn 79(083501):1–3

Saito G, Hosokai S, Akiyama T (2011a) Synthesis of ZnO nanoflowers by solution plasma. Mater Chem Phys 130:79–83

Saito G, Hosokai S, Tsubota M, Akiyama T (2011b) Synthesis of copper/copper oxide nanoparticles by solution plasma. J Appl Phys 110(023302):1–6

Saito G, Hosokai S, Tsubota M, Akiyama T (2012) Influence of solution temperature and surfactants on morphologies of tin oxide produced using a solution plasma technique. Cryst Growth des 12:2455–2459

Saito G, Azman WOSBWM, Nakasugi Y, Akiyama T (2014a) Optimization of electrolyte concentration and voltage for effective formation of Sn/SnO_2 nanoparticles by electrolysis in liquid. Adv Powder Technol 25:1038–1042

Saito G, Nakasugi Y, Yamashita T, Akiyama T (2014b) Solution plasma synthesis of bimmetallic nanoparticles. Nanotechnol 25(135603):1–7

Saito G, Nakasugi Y, Yamashita T, Akiyama T (2014c) Solution plasma synthesis of ZnO flowers and their photoluminescence properties. Appl Surf Sci 290:419–424

Saito G, Zhu C, Akiyama T (2014d) Surfactant-assisted synthesis of Sn nanoparticles via solution plasma technique. Adv Powder Technol 25:728–732

Sano N, Wang H, Chhowalla M, Alexandrou I, Amaratunga GAJ (2001) Synthesis of carbon 'onions' in water. Nature 414:506–507

Sano N, Wang H, Alexandrou I, Chhowalla M, Teo KBK, Amaratunga GAJ, Iimura K (2002) Properties of carbon onions produced by an arc discharge in water. J Appl Phys 92:2783–2788

Sato S, Mori K, Ariyada O, Atsushi H, Yonezawa T (2011) Synthesis of nanoparticles of silver and platinum by microwave-induced plasma in liquid. Surf Coat Technol 206:955–958

Schiros T, Nordlund D, Pa´lova´ L, Prezzi D, Zhao L, Kim KS, Wurstbauer U, Gutie´rrez C, Delongchamp D, Jaye C, Fischer D, Ogasawara H, Pettersson LGM, Reichman DR, Kim P, Hybertsen MS, Pasupathy AN (2012) Connecting dopant bond type with electronic structure in N-doped graphene. Nano Lett 12:4025–4031

Schwyn S, Garwin E, Schmidt-Ott A (1988) Aerosol generation by spark discharge. J Aerosol Sci 19:639–642

Shi J, Hu X, Zhang J, Tang W, Li H, Shen X, Saito N (2014) One-step facile synthesis of Pd nanoclusters supported on carbon and their electrochemical property. Prog Nat Sci Mater Int 24:593–598

Shirafuji T, Noguchi Y, Yamamoto T, Hieda J, Saito N, Takai O, Tsuchimoto A, Nojima K, Okabe Y (2013) Functionalization of multi-walled carbon nanotubes by solution plasma processing in ammonia aqueous solution and preparation of composite material with polyamide 6. Jpn J Appl Phys 52(125101):1–6

Shirai N, Uchida S, Tochikubo F (2014) Synthesis of metal nanoparticles by dual plasma electrolysis using atmospheric dc glow discharge in contact with liquid. Jpn J Appl Phys 53(046202):1–5

Sudare T, Ueno T, Watthanaphanit A, Saito N (2015a) Accelerated nanoparticles synthesis in alcohol-water-mixture-based solution plasma. Phys Chem Chem Phys 17:30255–30259

Sudare T, Ueno T, Watthanaphanit A, Saito N (2015b) Verification of radicals formation in ethanol-water mixture based solution plasma and their relation to the rate of reaction. J Phys Chem A 119:11668–11673

Šunka P, Babický V, Člupek M, Lukeš P, Šimek M, Schmidt J, Černák M (1999) Generation of chemically active species by electrical discharges in water. Plasma Sources Sci Technol 8:258–265

Tantiplapol T, Singsawat Y, Narongsil N, Damrongsakkul S, Saito N, Prasertsung I (2015) Influences of solution plasma conditions on degradation rate and properties of chitosan. Innov Food Sci Emerg Technol 32:116–120

Tao X, Sun C, Han Y, Huang L, Xu D (2019) The plasma assisted preparation of Fe-MOFs with high adsorption capacity. Cryst Eng Comm 21:2541–2550

Tarasenko N, Nevar A, Nedelko M (2010) Properties of zinc-oxide nanoparticles synthesized by electrical-discharge technique in liquids. Phys Status Solidi A 207:2319–2322

Terashima C, Iwai Y, Cho SP, Ueno T, Zettsu N, Saito N, Takai O (2013) Solution plasma sputtering processes for the synthesis of PtAu/C catalysts for Li-air batteries. Int J Electrochem Sci 8:5407–5420

Thongwichit N, Li O L H, Yaowarat W, Saito N, Suriyapraphadilok U (2016) Adsorption of carbon dioxide by solution-plasma-synthesized heteroatom-doped carbon nanospheres. Jpn J Appl Phys 55:01AE10(1–7pp)

Todaro M, Buscarino G, Sciortino L, Alessi A, Messina F, Taddei M, Ranocchiari M, Cannas M, Gelardi FM (2016) Decomposition process of carboxylate MOF HKUST-1 unveiled at the atomic scale level. J Phys Chem C 120:12879–12889

Tokushige M, Nishikiori T, Ito Y (2009a) Synthesis of Ni nanoparticles by plasma-induced cathodic discharge electrolysis. J Appl Electrochem 39:1665–1670

Tokushige M, Yamanaka T, Matsuura A, Nishikiori T, Ito Y (2009b) Synthesis of magnetic nanoparticles (Fe and FePt) by plasma-induced cathodic discharge electrolysis. IEEE Trans Plasma Sci 37:1156–1160

Tokushige M, Nishikiori T, Ito Y (2010a) Formation of fine Ni nanoparticle by plasma-induced cathodic discharge electrolysis using rotating disk anode. J Electrochem Soc 157:E162–E166

Tokushige M, Nishikiori T, Ito Y (2010b) Plasma-induced cathodic discharge electrolysis to form various metal /alloy nanoparticles. Russ J Electrochem 46:619–626

Tong DG, Luo YY, Chu W, Guo YC, Tian W (2010) Cutting of carbon nanotubes via solution plasma processing. Plasma Chem Plasma Process 30:897–905

Tong D, Wu P, Su P, Wang D, Tian H (2012) Preparation of zinc oxide nanospheres by solution plasma process and their optical property, photocatalytic and antibacterial activities. Mater Lett 70:94–97

Tong DG, Chu W, Wu P, Gu GF, Zhang L (2013) Retracted article: mesoporous multiwalled carbon nanotubes as supports for monodispersed iron-boron catalysts: improved hydrogen generation from hydrous hydrazine decomposition. J Mater Chem A 1:358–366

Torimoto T, Ki O, Kiyama T, Hirahara K, Tanaka N, Kuwabata S (2006) Sputter deposition onto ionic liquids: Simple and clean synthesis of highly dispersed ultrafinemetal nanoparticles. Appl Phys Lett 89(243117):1–3

Watthanaphanit A, Saito N (2013) Effect of polymer concentration on the depolymerization of sodium alginate by the solution plasma process. Polym Degrad Stab 98:1072–1080

Watthanaphanit A, Heo YK, Saito N (2014a) Influence of the discharge time of solution plasma process on the formation of gold nanoparticles in alginate matrix. J Taiwan Inst Chem Eng 45:3099–3103

Watthanaphanit A, Panomsuwan G, Saito N (2014b) A novel one-step synthesis of gold nanoparticles in an alginate gel matrix by solution plasma sputtering. RSC Adv 4:1622–1629

Wu P, Sun J, Huang Y, Gu G, Tong D (2012) Solution plasma synthesized nickel oxide nanoflowers: an effective NO_2 sensor. Mater Lett 82:191–194

Xie YB, Liu CJ (2008) Stability of ionic liquids under the influence of glow discharge plasmas. Plasma Process Polym 5:239–245

Xie SY, Ma ZJ, Wang CF, Lin SC, Jiang ZY, Huang RB, Zheng LS (2004) Preparation and self-assembly of copper nanoparticles via discharge of copper rod electrodes in a surfactant solution: a combination of physical and chemical processes. J Solid State Chem 177:3743–3747

Xu K, Fu Y, Zhou Y, Hennersdorf F, Machata P, Vincon I, Weigand JJ, Popov AA, Berger R, Feng X (2017) Cationic nitrogen-doped helical nanographenes. Angew Chem Int Ed 56:15876–15881

Yadav R, Dixit C (2017) Synthesis, characterization and prospective applications of nitrogen-doped graphene: a short review. J Sci: Adv Mater Devices 2:141–149

Yamamoto M, Yoshida T, Yamamoto N, Nomoto T, Yagi S (2015) The loading effect of silver nanoparticles prepared by impregnation and solution plasma methods on the photocatalysis of Ga_2O_3. Nucl Instrum Methods Phys Res B 359:64–68

Yaowarat W, Li OLH, Saito N (2016) Highly durable silica-coated Pt/carbon nanotubes for proton-exchange membrane fuel cells application. Jpn J Appl Phys 55:01AE23(1–5pp)

Yun KS, Kim BR, Kang WS, Jung SC, Myung ST, Kim SJ (2013) Preparation of carbon blacks by liquid phase plasma (LPP) process. J Nanosci Nanotechnol 13:7381–7385

Yun K, Kim B, Noh E, Jung H, Jung S, Kang W, Myung S, Kim S (2014a) Effect of water on the performance of carbon blacks anode material processed by plasma in the benzene solution. Sci Adv Mater 6:1594–1598

Yun KS, Kim BR, Kim SC, Jung SC, Kang WS, Kim SJ (2014b) Carbon black nanoparticles with a high reversible capacity synthesized by liquid phase plasma process. Res Chem Intermed 40:2559–2564

Yun KS, Kim BR, Noh E, Jung HJ, Oh HJ, Kang W, Jung SC, Myung ST, Kim SJ (2014c) Performance improvement of liquid phase plasma processed carbon blacks electrode in lithium ion battery applications. Int J Precis Eng Manuf 15:1689–1693

Zhang ZK, Bai ML, Guo DZ, Hou SM, Zhang GM (2011) Plasma-electrolysis synthesis of TiO_2 nano/microspheres with optical absorption extended into the infra-red region. Chem Commun 47:8439–8441

Zhang J, Hu X, Shi J, Lu P, Shen X, Xu P, Saito N (2015) Synthesis of SnO_2 nanoparticles using a solution plasma and their gas-sensing properties. Jpn J Appl Phys 55(1S):01AE17(1–4pp)

Zhu H, Li X, Jiang B, Xu C, Zhu Y, Wu D, Chen X (2002) Formation of carbon nanotubes in water by the electric-arc technique. Chem Phys Lett 366:664–669

Zhu Q, Huang X, Zeng Y, Sun K, Zhou L, Liu Y, Luo L, Tian S, Sun X (2021) Controllable synthesis and electrocatalytic applications of atomically precise gold nanoclusters. Nanoscale Adv 3:6330–6341

Toward Industrial Solution Plasma Process

Nagahiro Saito and Tatsuru Shirafuji

Abstract Solution plasma processes (SPPs) possess high potential for applications such as nanomaterial synthesis. However, challenges remain in scaling them up to industrial levels. This chapter outlines advancements in reactor design and power supply technology aimed at achieving such scalability.

1 Introduction

As described in the previous chapters of this book, solution plasma processes (SPPs) have the possibility to be applied to various fascinating applications, especially in the synthesis of nanomaterials. These studies have been conducted with small laboratory-level experimental equipment. In such equipment, plasma is generated in the gap (approx. 1 mm) between a pair of electrodes (diam. approx. 1 mm). The size of plasma is, hence, the mm scale. This size of plasma is quite smaller than the volume of the liquid to be treated, even if the liquid is as small as a few 10 mL. Since this solution plasma looks like a dot in the liquid, we call it zero-dimensional SP in this chapter.

In the case of industrial application, the volume of liquid to be treated can be a few 10 L a day. Thus, if you apply SPPs to an industrial application, you must develop a suitable reactor for processing large amounts of liquid. This chapter describes several efforts toward industrial SPPs.

N. Saito (✉)
Department of Chemical Systems Engineering, Graduate School of Engineering, Institute for Innovation for Future Society, Nagoya University, Furocho, Chikusa, Nagoya, Japan
e-mail: saito.nagahiro.z7@f.mail.nagoya-u.ac.jp

T. Shirafuji
Department of Physics and Electronics, Graduate School of Engineering, Osaka Metropolitan University, Nakaku, Sakai, Osaka, Japan
e-mail: shirafuji@omu.ac.jp

N. Saito (ed.), *Solution Plasma Chemistry*, Lecture Notes in Chemistry 115,
https://doi.org/10.1007/978-981-95-0966-9_7

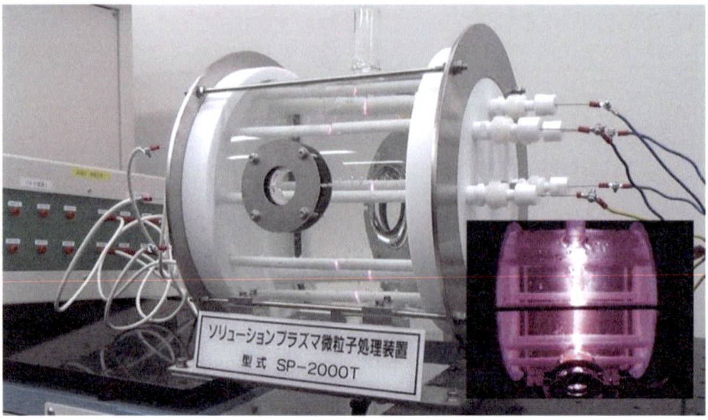

Fig. 1 A multi-electrode system (barrel type with 8 parallel electrodes)

2 Multi-electrode System

The first approach to achieving SPP for processing large volumes of liquid was the use of multiple electrodes, as shown in Fig. 1, which has 8 electrode pairs. By using this system, you can process liquids with a larger volume than in a small conventional reactor with a single pair of electrodes. However, as you notice, the plasma between the electrodes is still tiny. Since the direct effect from the plasma can extend only to the vicinity of the electrodes, it is necessary to stir or flow the liquid in the reactor or to move the electrodes to treat the liquid being far away from the plasma regions.

Figure 2a, b shows a prototype reactor with a gravitational liquid flow system. Multiple processing is possible by making multiple stages, as shown in the figure. Figure 2c, d shows a reactor with the function of reciprocating the electrodes in the XY direction. If you need more efficient processes, you can add the function of reciprocating electrodes along the Z-axis. These two methods can be combined for more efficient treatment processes.

3 Pulse Power Sources with High Frequency

Regarding the pulse power source for SPPs, its frequency is one of the important factors since the pulse frequency determines the number of discharges per unit time, and the higher the frequency, the higher the processing speed. When the solution plasma was first realized, the pulse frequency was at most 15 kHz. Subsequent improvements in power supply technology improved the pulse frequency to 30 kHz. Figure 3 shows the change in the intensity of the emission spectrum with such an increase in pulse frequency. It can be seen that when the pulse frequency of the applied voltage doubles, the emission intensity of the plasma also doubles.

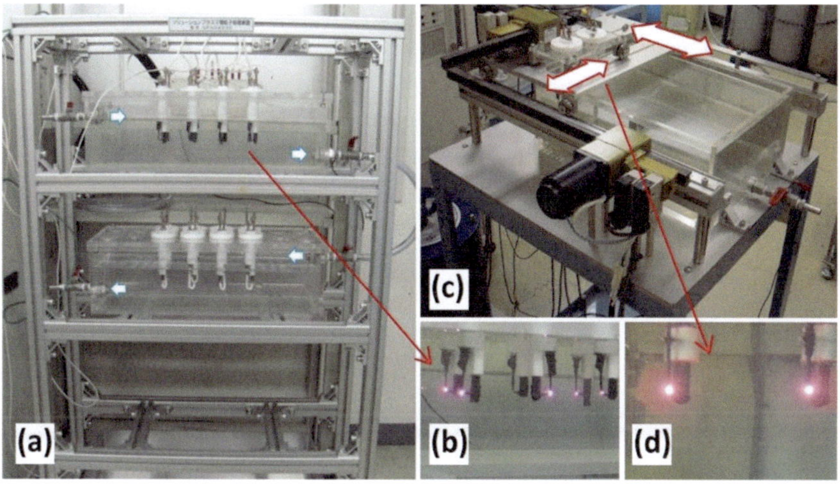

Fig. 2 **a** A multi-electrode system with a tandem flow system, and **b** a close-up photograph of the implemented electrodes. **c** A multi-electrode system with an X–Y scanning system, and **d** a close-up photograph of the implemented electrodes

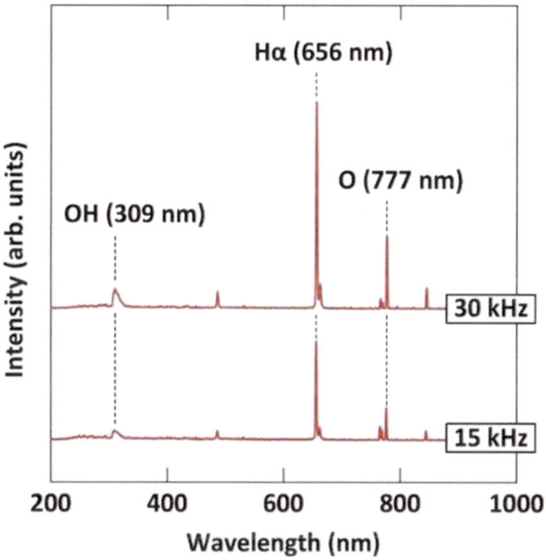

Fig. 3 Optical emission spectra of the SPs operated with pulse frequencies of 15 and 30 kHz. The higher the frequency, the higher the emission intensity

What power supply should you choose (or build) for SPP at higher frequency pulse voltages? The upper limit of the pulse frequency depends on the properties of the semiconductor material used for the inverter devices. Conventionally, silicon (Si) has been used as a semiconductor material for switching devices. Recently, switching devices using silicon carbide (SiC), which is known as a wide-gap semiconductor, have been put into practical use (Matsunami 2004; Kimoto and Cooper 2014; Kimoto and Watanabe 2020). Thanks to the recent development of SiC technology, a pulse power supply with a frequency beyond 1 M kHz is also available (Okuda et al. 2021). SiC has another advantage of low "on-resistance" (Baliga 2005, 2019; Yonezawa et al. 2018). In the case of conventional Si inverters, a considerable amount of electric power is lost as thermal energy due to Joule heating of Si devices when they are in the "on" state. In the case of SiC inverters, thanks to the low "on-resistance" characteristics, they can generate pulse power with less heating. Thus, using SiC inverters, you can perform SPPs with high energy efficiency.

3.1 A Note on Parallel Discharge

The multi-electrodes system is a very simple idea and easy to implement in the conventional reactor system. However, one must pay attention to the behavior of electrical discharges or plasma generation and basic theory in electric circuits in the case of multiple parallel electrodes. In theory, when we apply voltage on parallel circuits, the same voltage is applied to all the parallel paths, which correspond to the electrode pairs in the reactor. If the voltage is not high enough for igniting discharge between these electrode pairs, you must increase the applied voltage. Then, you will increase the voltage to ignite discharges. Maybe you expect bright plasma to appear simultaneously between all pairs of electrodes when the voltage becomes high enough to ignite discharge.

However, if the discharge is once ignited between only one of these electrodes, you do not have what you expect. Since plasma is a conductor, the gap between the electrodes is almost short-circuited when the discharge is ignited between them. This means that the voltage between these electrodes will drop rapidly. In the situation, after a discharge occurs between the electrodes, the discharge is maintained even if the voltage drops after that because the threshold voltage for igniting discharge is generally higher than that for sustaining discharge. However, please note that discharges are not yet ignited between the other electrode pairs at that moment. Additionally, the voltage drop also occurs for the other electrode pairs because they are connected in parallel. If the voltage drops below the ignition threshold voltage before starting discharge, you cannot expect discharges anymore between the other electrodes. Consequently, only one plasma is generated between one of the electrode pairs.

The simple method to overcome this issue is to use separate power sources for each electrode pair. However, this idea is not cost-effective. The alternative method is to connect a capacitor to each electrode pair in series since the charged capacitor works as an effective power source that can apply voltage independently of the power supply, although the capacitance should be high enough to keep high voltage during ignition of the pulse discharge.

4 Narrow Channel but Large Volume

Figures 1 and 2c are so-called batch reactors in which the volume of the reactor limits the total amount of liquid to be processed. If you want to process more liquid, it is necessary to prepare a reactor with a larger volume or multiple reactors. The other approach is to use a flow system, which may be more cost-effective because a large amount of liquid can be processed with a small-volume reactor. Although such a system is already shown in Fig. 2b, there is a massive dead volume where the liquid does not make direct contact with the plasma. To improve this situation, the liquid flow channel should be narrower.

Figure 4 shows a prototype of the flow-type reactor with narrower flow channels. A similar system shown in Fig. 5 has been used in industry. In theory, only one pair of electrodes is sufficient when a liquid is circulated. However, since the plasma is still tiny compared to the diameter of the flow channels, most of the flowing liquid still passes through the electrode region without direct contact with the plasma. For industrial applications, it is better to have as little untreated liquid as possible when the liquid is circulating the flow channel. Therefore, multiple electrodes are along the flow path in these reactors.

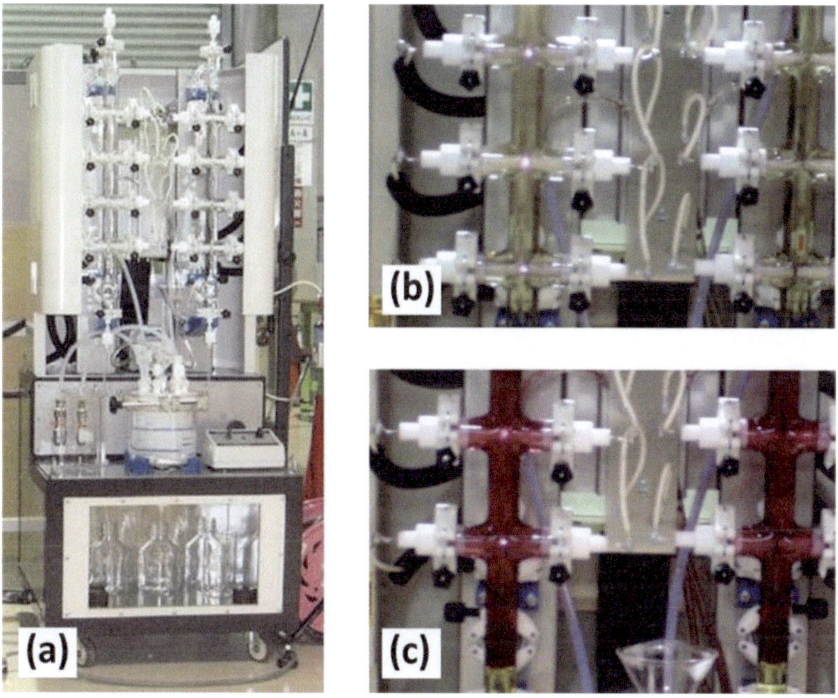

Fig. 4 Multi-electrode system (flow type). **a** Overview, **b** just after discharge ignition and **c** after continuous discharge for 60 min in $HAuCl_4$ aqueous solution

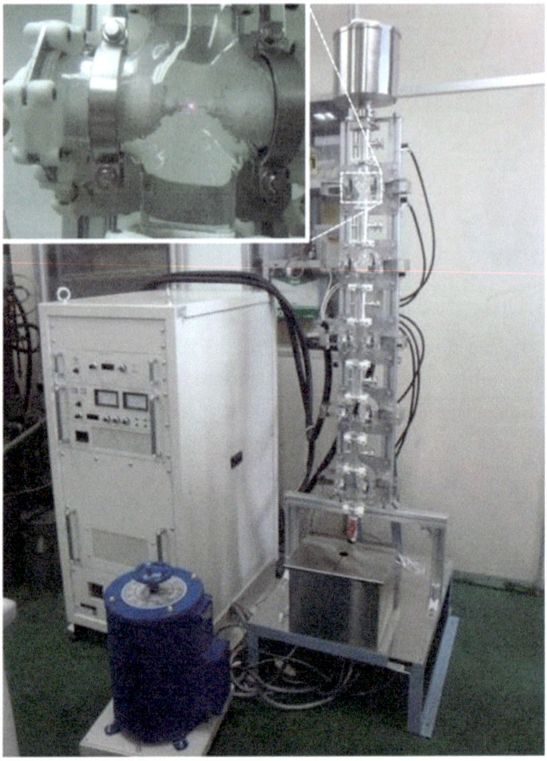

Fig. 5 A multi-electrode system for treatment of MWCNT, which has been used in industry. Photos by courtesy of Toyo Jushi Corporation

5 Two-Dimensionally Distributed Microplasmas

In the previous section, the method to scale up an SPP reactor was to simply increase the number of discrete electrodes. As long as discrete electrodes are used, however, there will always be a liquid that does not come into direct contact with the plasma. Ideally, it is desirable to create a situation in which numerous microplasmas are in contact with the liquid because what we should enlarge is not plasma volume but plasma/liquid interface. This section describes several attempts to realize this basic idea.

The first one is the reactor in which a two-dimensionl distributed microplasma is generated at the interface of a dielectric plate and porous dielectric material containing a liquid medium. Figure 6a shows the experimental setup for realizing the 2D distributed microplasmas. We employed Bascotect (BASF) as the porous dielectric material, which is made of melamine resin and has an open-cell foam structure with a pore size of 200–500 μm. The aqueous solution to be treated is held in the porous dielectric foam to which the quartz plate is attached. At the interface between the quartz plate and porous dielectric foam, tiny gas gaps are formed, as shown in Fig. 6b.

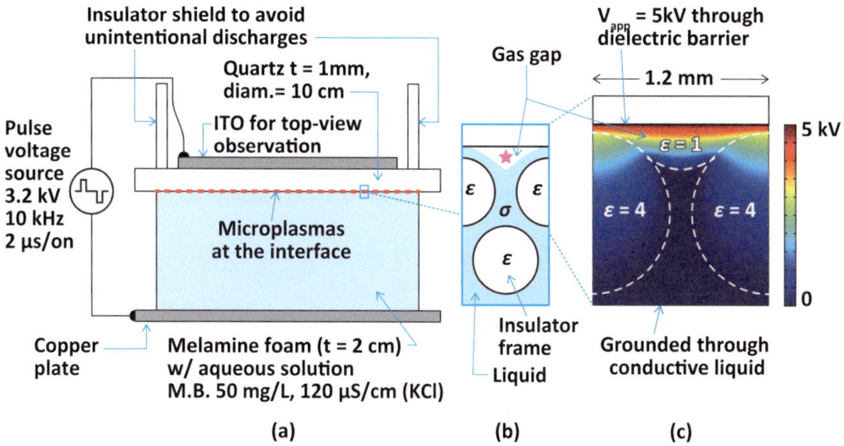

Fig. 6 **a** Schematic drawings of a reactor, which can generate two-dimensionally distributed microplasmas at the interface between dielectric plate and porous dielectric material, **b** microscopic view near a gas-gap formed between the dielectric plate and porous dielectric material filled with conductive liquid, and **c** potential profile calculated around the gas-gap

Each tiny gas-gap region constructs the structure of quartz/gas-gap/conductive aqueous solution, which forms a configuration for dielectric barrier discharge. The top side of the quartz plate is coated with a thin film of transparent conductive oxide (TCO) made of indium tin oxide (diameter 10 cm) for applying the voltage on it and also for observing plasma at the interface through the quartz plate. Figure 6c shows a numerical simulation result of the potential profile around the gas-gap region under the condition that high voltage is applied between the TCO electrode on the quartz plate and the bottom electrode (a copper plate) below the porous dielectric foam. We can see a significant potential drop in the gas gap. In this work, the voltage applied between the electrodes is pulsed voltage with an amplitude of 5 kV, frequency of 10 kHz, and width of 2 μs. The conductivity of the aqueous solution held in the porous foam is controlled to be 120 μS/cm using KCl.

Figure 7 shows an optical image when a pulse voltage is applied between the electrodes. Many microplasmas are formed at the interface between the quartz plate and porous foam. The actual size of the microplasma is approximately 300 μm, which corresponds to the pore size of the porous melamine foam employed in this work. Conventional SPPs also employ tiny plasma in contact with the liquid, but its number is only a few at most. In this work, the size of each microplasma is as small as 300 μm, but their number is almost uncountable, and their integrated size is as large as 10 cm in diameter. Furthermore, scaling toward larger diameters is possible if the voltage source has sufficient current capacity.

This reactor has been applied for the decomposition of methylene blue in an aqueous solution (Shirafuji et al. 2011). Figure 8 shows UV–VIS spectra of the methylene blue aqueous solution treated with this reactor. We can see the decrease

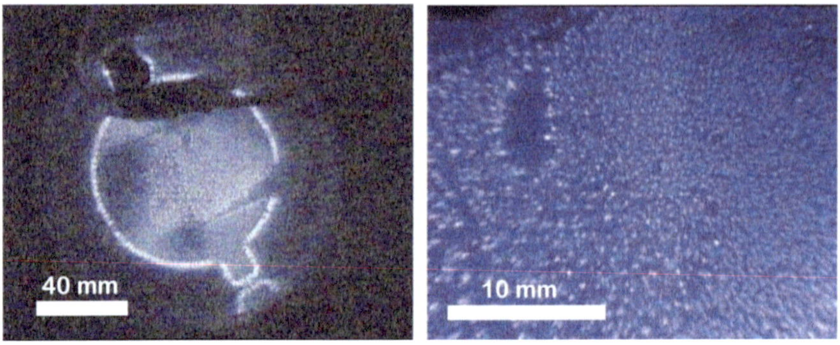

Fig. 7 The two-dimensionally distributed microplasmas observed through the top quartz electrode

Fig. 8 UV–VIS spectra of the methylene blue aqueous solution treated with a 2D distributed microplasma reactor

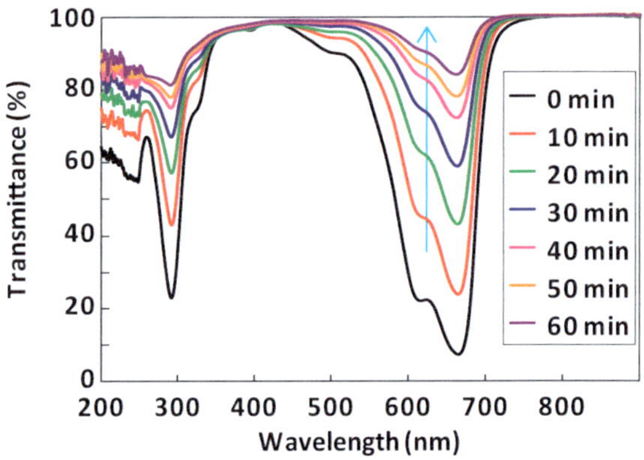

in the absorption band at 600–700 nm, which arises due to optical absorption by methylene blue molecules in the aqueous solution.

6 Three-Dimensionally Distributed Microplasmas

Although the idea of two-dimensionally distributed microplasmas is proposed in the previous subsection, this idea still has the same problem as the zero-dimensional SPPs. The problem is the fact that only the liquid near the surface of the porous material comes into contact with the plasma. Ideally, the favored scheme is to generate many microplasmas which are three-dimensionally dispersing in the liquid or those inside the porous material involving liquid. This subsection describes the concept and realization of three-dimensional distributed microplasmas in liquid.

Figure 9 shows the transition of ideas for increasing the efficiency and capacity of SPPs. Figure 9a indicates a zero-dimensional SP, which corresponds to the original

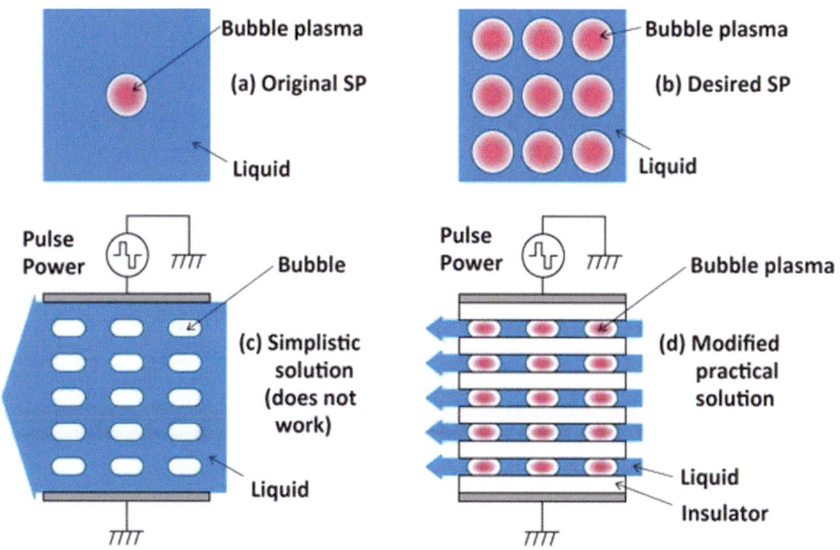

Fig. 9 The basic idea for increasing the efficiency and capacity of SPPs

SPP system. Figure 9b shows the fundamental idea of enlarging the interface between plasma and liquid, in which it is desirable that the microplasmas generated in the liquid are as small as possible, abundant, and uniformly dispersed in the liquid.

Figure 9c is an idea according to simplistic thinking, in which simple bubbling of the liquid is assumed to be good enough for achieving the situation in Fig. 9b. However, Horii (1973) clarified that plasma could not be generated by the method shown in Fig. 9c. Horii attempted in-bubble discharges in the experiment shown in Fig. 10a. However, he could not achieve in-bubble discharges in the case of Fig. 10a, in which bubbles are freely suspended in the liquid. Alternatively, in the case shown in Fig. 10b, where the bubbles are in contact with the electrodes, the discharge occurred.

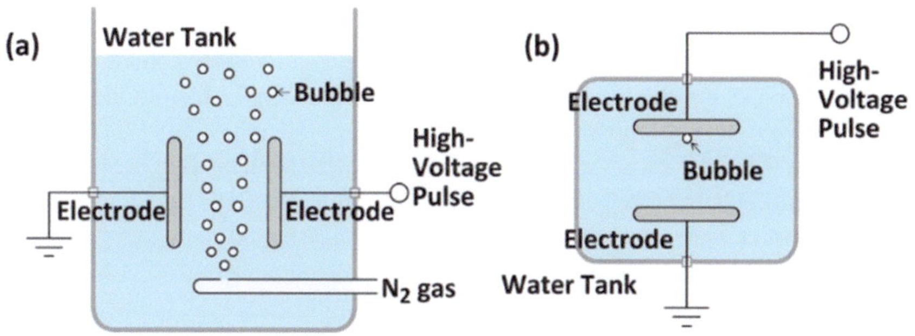

Fig. 10 Schematic illustrations of Horii's experiments (Horii 1973)

Note that when it was irradiated with X-rays, the discharge occurred even in the case of Fig. 10a. From these experimental results, Horii concluded that discharges would not occur in the bubbles fully surrounded by the liquid because the number of initial electrons required to start the discharge was extremely small. He also deduced the conclusion that discharge occurs when X-ray irradiation that generates numerous initial electrons is performed or when the bubbles are in contact with a solid surface where initial electrons can be emitted. Figure 9d is a modification of Fig. 9c based on Horii's conclusion, which is characterized by the fact that bubbles are in contact with the solid surface.

Figure 11 is a pseudo-realization of the system of Fig. 9d (Shirafuji et al. 2014). In this method, a continuous porous dielectric, like a trabecular bone structure, is embedded in a glass tube. A metal electrode that supplies gas (Ar) and applies voltage is provided in the central axis of the porous dielectric, and a mesh-shaped ground electrode wraps the outer wall of the glass tube. In the continuous porous material, numerous microplasmas in fine bubbles are transported together with the liquid. The liquid is affected by the microplasmas during their transport. We call such solution plasma three-dimensional integrated micro solution plasma (3-D IMSP). To continuously discharge with 3-D IMSP, it is necessary to apply a voltage with a bipolar rectangular pulse waveform with a short rise time due to its electrical circuit characteristics (Shirafuji et al. 2014). So far, material synthesis using 3-D IMSP has been reported on the synthesis of gold nanoparticles (Fig. 12) (Shirafuji et al. 2013) and hydrogen peroxide (Shirafuji et al. 2019).

Regarding the efficiency of 3-D IMSP, comparative evaluation in material synthesis has not yet been sufficiently performed, but the decomposition of raw materials is first necessary during synthesis. Therefore, the power efficiency of decomposition (decomposition amount per unit power amount) was compared with the conventional 0-D SP. The decomposition target material was methylene blue. Figure 13 shows UV–Vis absorption spectra of the methylene blue aqueous solution as a function of treatment time.

The power efficiency for the decomposition of methylene blue was calculated from the reduction $\Delta\alpha$ in the absorption coefficient and consumed electric energy P, which is calculated from the integration of the product of voltage and current waveforms during the treatment. Table 1 shows the comparison between the power efficiency of methylene blue decomposition by the 3-D IMSP and a conventional 0-D SP, which shows the power efficiency of the 3-D IMSP is 16-fold higher than that of the 0-D SP (Shirafuji and Himeno 2013).

Note that 0-D PS does not require a gas supply since steam bubbles are generated by Joule heating between electrode pairs. 3-D IMSP uses Ar gas (including a trace amount of water vapor from the gas–liquid interface) as the discharge gas.

Therefore, we also compared the cost efficiency (decomposition amount per unit cost), including the gas cost. As a result, it was found that 3-D IMSP is about fivefold more cost-effective than 0-D SP, even if the gas cost is taken into consideration.

Additionally, since 3-D IMSP can be used in-line (can be inserted into the liquid flow path), it can handle large volumes of liquid processing by continuous processing, even if the device itself is small. Furthermore, as in the case of chemical engineering,

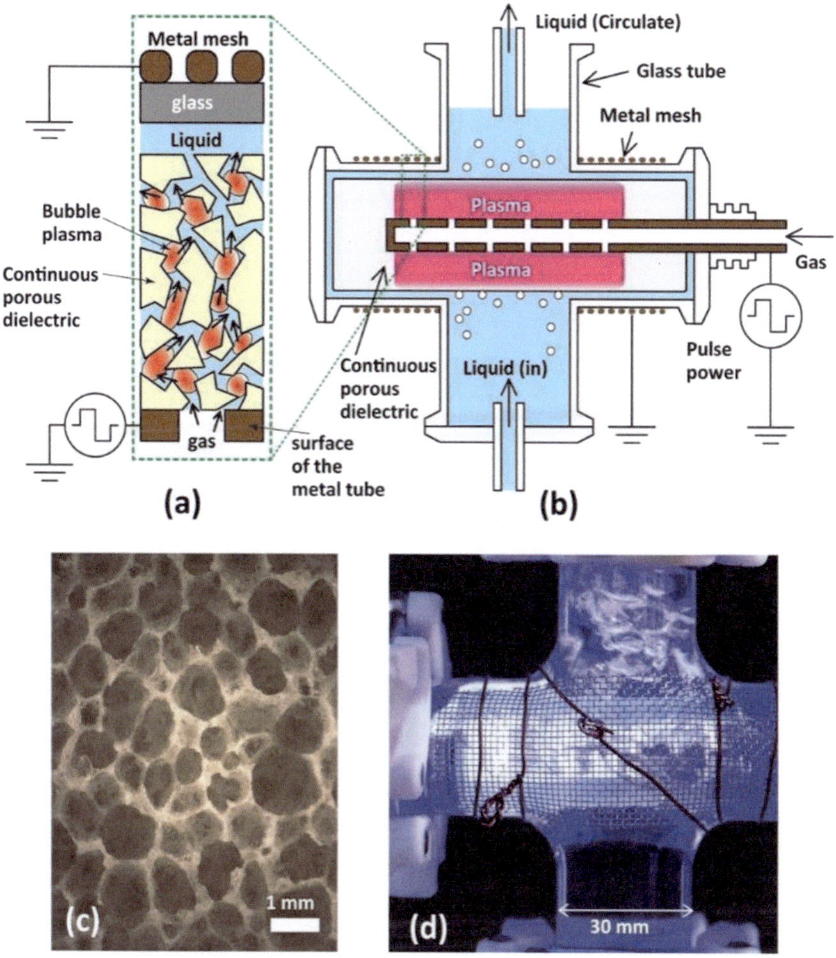

Fig. 11 **a** Diagram of plasma generation in fine bubbles moving through a continuous porous dielectric inside a 3-D IMSP device. **b** Schematic cross-sectional view of the 3-D IMSP device. **c** Overview of a continuous porous dielectric. **d** Overview of the 3-D IMSP device during discharge

the in-line use of multiple 3-D IMSP may be used for in-line multi-step processing, as shown in Fig. 14.

7 A Gas–Liquid Segmented Flow System

If you look into the 3-D IMSP, you will find gas (plasma)-liquid segmented flows, which are connected in series and parallel. Such a system is also seen in micro-chemistry (Köhler and Cahill 2014; Noël and Luque 2020; Buglioni et al. 2022)

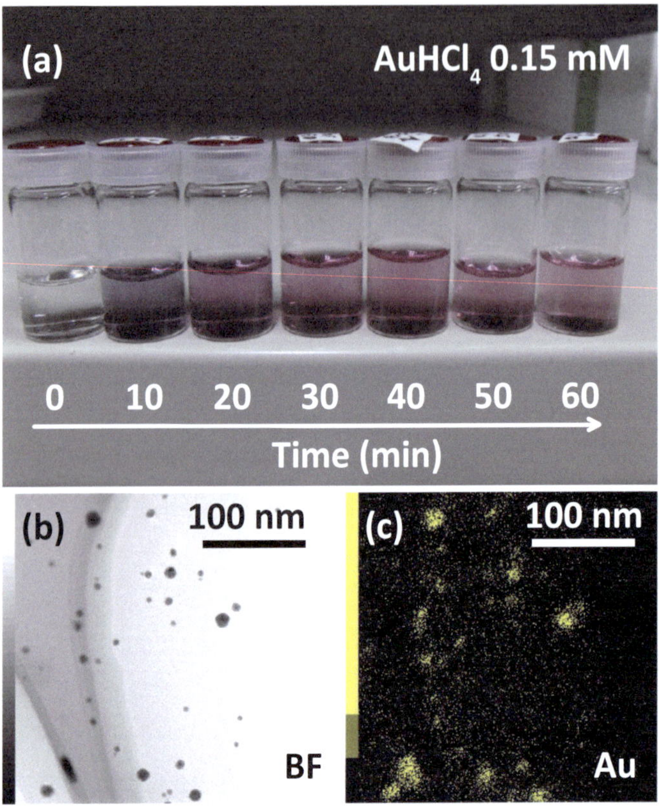

Fig. 12 Results of gold nanoparticle synthesis using 3D IMSP treatment of $HAuCL_4$ aqueous solution. **a** Aqueous solutions of $HAuCl_4$ after 3D IMSP treatment, **b** a STEM image of the synthesized particles, and **c** an EDS mapping image of gold in the STEM-observation area

and is important for process intensification flow chemistry (Han et al. 2020; Liu et al. 2020). In conventional gas–liquid segmented flows, the gas phase contains only stable molecules that are not so active. In some cases, UV-light irradiation is employed for the ignition of photolysis of these molecules to produce active species that will undergo reactions at the gas/liquid interfaces. However, UV photochemistry requires a UV-transparent material to construct a flow channel (Noël 2017). Generally, it is quartz that has high transmittance at UV wavelengths. Additionally, reaction products on the inner surfaces of the flow channels significantly reduce UV-light transmittance. Thus, you cannot perform a long operation using segmented microflow chemistry employing UV photolysis in some cases.

If you can use plasma chemistry instead of photochemistry, the above concerns will be dispelled. Is it possible? The answer is yes. Figure 15 shows a demonstration of plasma generation in a flowing segmented flow in a glass tube. Here, for simplicity, voltage is applied between the terminals of the tube. Alternatively, you can implement

Fig. 13 UV–VIS spectra of methylene blue aqueous solution treated with 3D IMSP

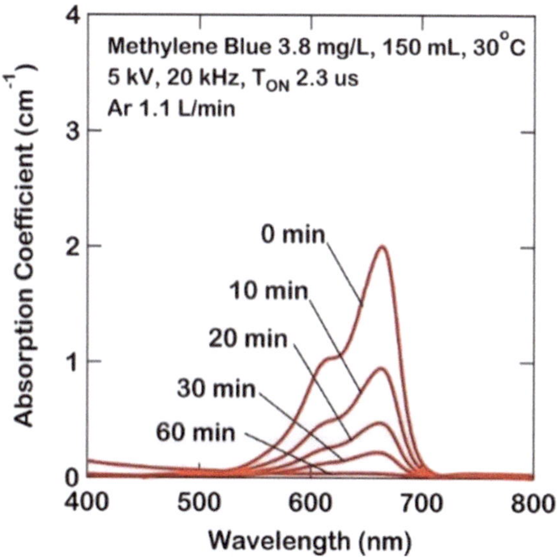

Table 1 Reduction in absorption coefficient $\Delta\alpha$ during the first 10 min of treatment, consumed electrical energy $\int P \cdot dt$ for the first 10 min, and efficiency η

Type	$\Delta\alpha$ (cm^{-1})	$E = \int P \cdot dt$ (Wh)	$\eta = \Delta\alpha/E$ (cm^{-1} W^{-1} h^{-1})
0-D SP	0.16	15.1	1.1×10^{-2}
3-D IMSP	1.06	6.75	15.7×10^{-2}

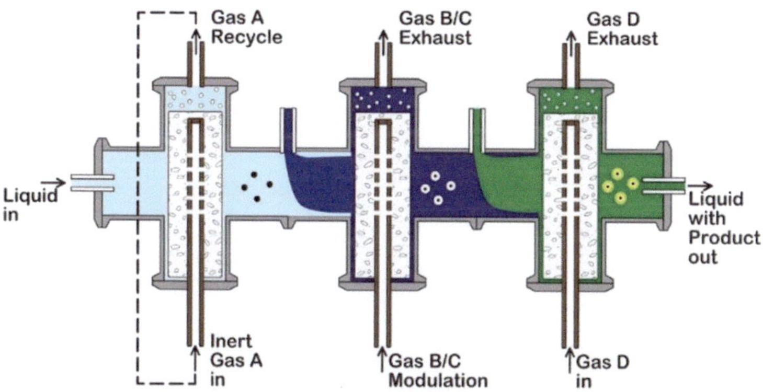

Fig. 14 Conceptual illustration of the possibility of in-line multi-step processing of nanomaterials using tandem 3-D IMSP devices, of which each 3-D IMSP plays a different role such as nuclei formation, growth, and encapsulation (for core–shell structure, for example)

segmented electrodes along the tube if you need. In both methods, one can make a reactor with bundled tubes without a reduction in plasma generation efficiency, while, in photochemistry, UV-light intensity significantly reduced in the central region of the bundled tubes.

Additionally, the combination of the concepts of the SPP and segmented flow is quite reasonable. The reason is found in Fig. 16. The most important region in an

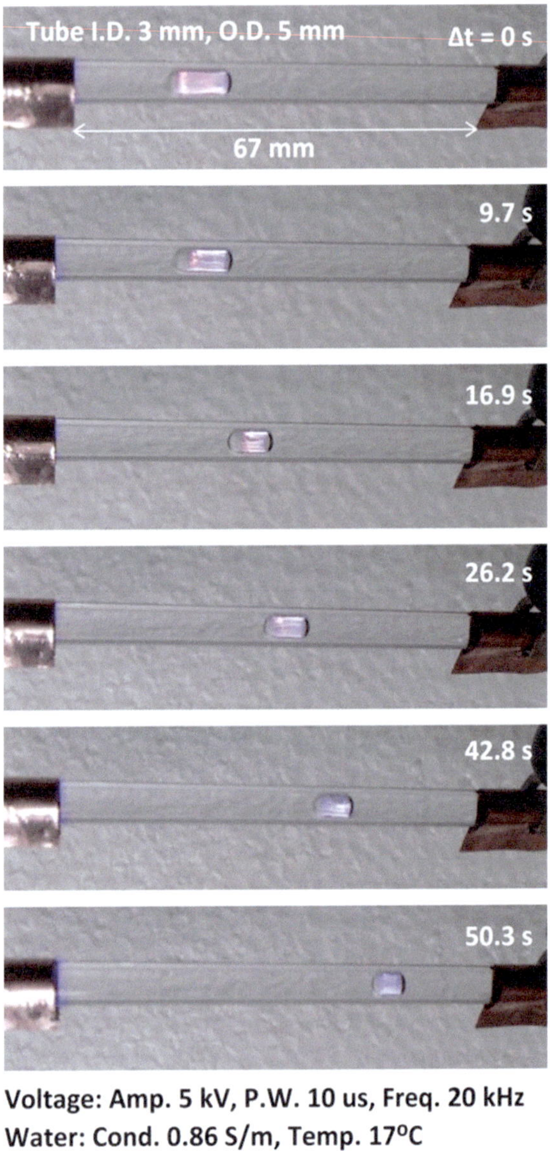

Fig. 15 An example of plasma generated in a gas phase in a simplified segmented flow

SPP is at the plasma-liquid interface, especially on the liquid side, where very active species originating from plasma trigger solution chemistry. However, the direct influence from plasma cannot reach deep inside the liquid due to the very high activity of the plasma-originated species. Thus, genuine plasma-driven reactions are localized at a very thin layer beneath the plasma-liquid interface. However, in a liquid phase in the segmented flow, there is continuous convection due to liquid flow. This phenomenon has the effects of pushing reaction products out of the interface region and dragging unreacted substances into the interface. This effect leads to the facilitation of the reactions to obtain the desired product(s). At present, there are few reports on such an approach. However, the combination of SPP and microflow chemistry may become one of the breakthrough technologies for the practical application of SPPs.

8 Summary

SPPs provide new material processing methods that bring plasma to the fields of electrochemical and solution chemistry and have the potential for various applications. To socially implement this new technology, it is essential to develop equipment for mass production. For this purpose, we focused on the fact that SPPs have the feature that "the region where the action of plasma directly affects is localized near the interface between plasma and liquid." In this chapter, we introduced 3-D IMSP for large-capacity liquid processing based on the idea of "massive plasma/liquid interfaces" as an example of making the best use of its features. The method described in this chapter has not yet been sophisticated. For example, the flow channels used in this 3-D IMSP are randomly distributed. If you use 3D printing technology, you can create designed flow channels, which may bring about much more efficient SPPs in the future.

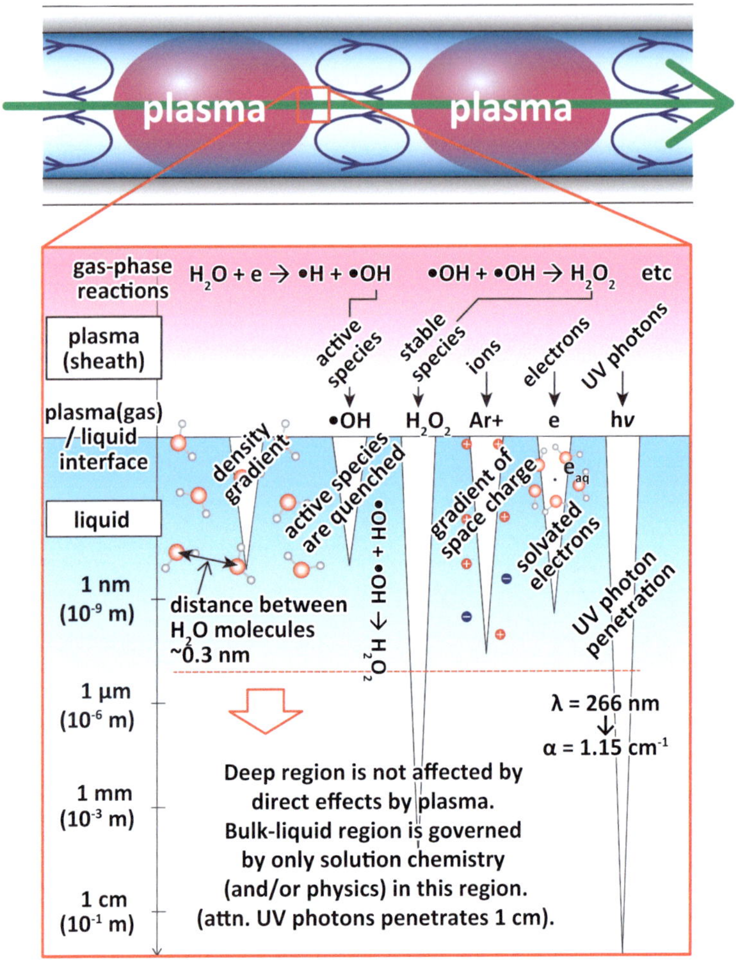

Fig. 16 A schematic illustration of plasma-liquid segmented flow and a conceptual image of plasma-liquid interface

References

Baliga BJ (2005) Silicon carbide power devices. World Scientific, Singapore

Baliga BJ (2019) Fundamentals of power semiconductor devices, 2nd edn. Springer, Switzerland

Buglioni L, Raymenants F, Slattery A, Zondag SDA, Noël T (2022) Technological innovations in photochemistry for organic synthesis: flow chemistry, high-throughput experimentation, scale-up, and photoelectrochemistry. Chem Rev 122:2752–2906

Han S, Kashfipour MA, Ramezani M, Abolhasani M (2020) Accelerating gas–liquid chemical reactions in flow. Chem Commun 56:10593–10606

Horii K (1973) Generation and application of discharges in bubbles in water. Denkigakkai Houdenkenkyukai Shiryo, ED-73-22:1–5 (in Japanese)

Kimoto T, Cooper JA (2014) Fundamentals of silicon carbide technology. Wiley, Singapore

Kimoto T, Watanabe H (2020) Defect engineering in SiC technology for high-voltage power devices. Appl Phys Express 13(12):120101
Köhler JM, Cahill BP (eds) (2014) Micro-segmented flow. Springer, Berlin
Liu Y, Chen G, Yue J (2020) Manipulation of gas–liquid–liquid systems in continuous flow microreactors for efficient reaction processes. J Flow Chem 10:103–121
Matsunami H (2004) Technological breakthroughs in growth control of silicon carbide for high power electronic devices. Jpn J Appl Phys 43(10):6835–6847
Noël T (ed) (2017) Photochemical processes in continuous-flow reactors. World Scientific, Singapore
Noël T, Luque R (eds) (2020) Accounts on sustainable flow chemistry. Springer, Switzerland
Okuda T, Nishimura Y, Nishioka K, Kishimoto S, Kikuchi Y, Nakamura T (2021) A 5-kV pulse generator with a 100-kV/μs slew rate based on series-connected 1700-V SiC MOSFETs for electrical insulation tests. Rev Sci Instrum 92:114705
Shirafuji T, Himeno Y (2013) Generation of three-dimensionally integrated micro-solution plasmas and its application to decomposition of methylene blue molecules in water. Jpn J Appl Phys 52(11):11NE03
Shirafuji T, Hieda J, Saito N, Takai O (2011) Application of integrated micro solution plasma to decomposition of organic substances in water. In: 20th International symposium on plasma chemistry. Philadelphia, PA, ID640, pp 1–2
Shirafuji T, Ueda J, Nakamura A, Cho SP, Saito N, Takai O (2013) Gold nanoparticles synthesis by using three-dimensionally integrated micro solution plasmas. Jpn J Appl Phys 52(12):126202
Shirafuji T, Nomura A, Himeno Y (2014) Three-dimensionally integrated micro-solution plasma: numerical feasibility study and practical applications. Plasma Chem Plasma Process 34(3):523–534
Shirafuji T, Miyamoto S, Oh JS (2019). H_2O_2 synthesis using 3D integrated micro solution plasma and improvement of its efficiency by combining with ion exchange resin. In: Materials research meeting 2019, December 10–14. Yokohama, H1-11-I03
Yonezawa Y, Nakayama K, Kosugi R, Harada S, Koseki K, Sakamoto K, Kimoto T, Okumura H (2018) Progress in high and ultrahigh voltage silicon carbide device technology. In: 2018 IEEE International electron devices meeting (IEDM). San Francisco, CA, December 1–5, pp 19.3.1–19.3.4

Index

N. Saito (ed.), *Solution Plasma Chemistry*, Lecture Notes in Chemistry 115,
https://doi.org/10.1007/978-981-95-0966-9

MIX
Papier aus verantwortungsvollen Quellen
Paper from responsible sources
FSC® C105338

If you have any concerns about our products,
you can contact us on
ProductSafety@springernature.com

In case Publisher is established outside the EU,
the EU authorized representative is:
Springer Nature Customer Service Center GmbH
Europaplatz 3, 69115 Heidelberg, Germany

Printed by Libri Plureos GmbH
in Hamburg, Germany